math
a
day

by theoni pappas

Wide World Publishing/Tetra

Wide World Publishing/Tetra
P.O. Box 476
San Carlos, CA 94070

websites:
http://www.wideworldpublishing.com
http://www.mathproductsplus.com

Printed in the United States of America.

3rd Printing May 2004
Revised

ISBN: 1-884550-20-7

Library of Congress Cataloging-in-Publication Data

Pappas, Theoni
   Math-a-day : a book of days for your mathematical year / Theoni Pappas.
      p. cm.
   Includes index.
   ISBN 1-884550-20-7
   1. Mathematics-- Miscellanea. I. Title

QA99 .P375 1999
510--dc21   99-088313

—Table of Contents—

Each day has
- its own math quote
- its own problem or puzzle
- its own historical or current note
- the date written in two number systems

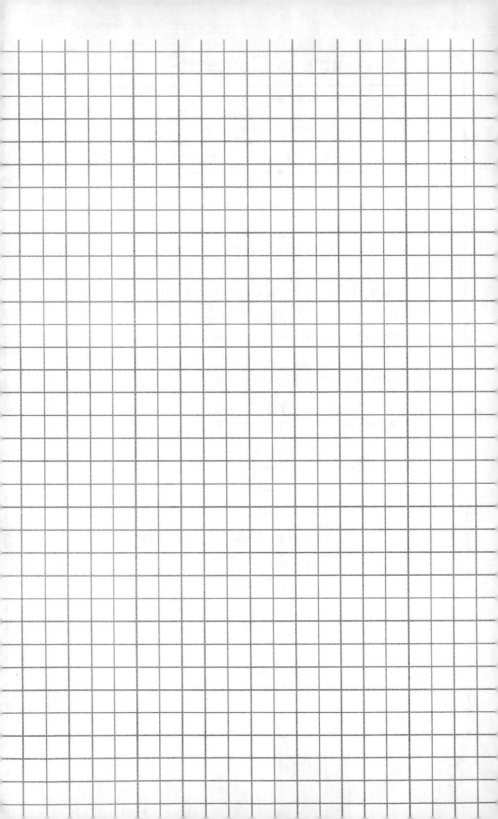

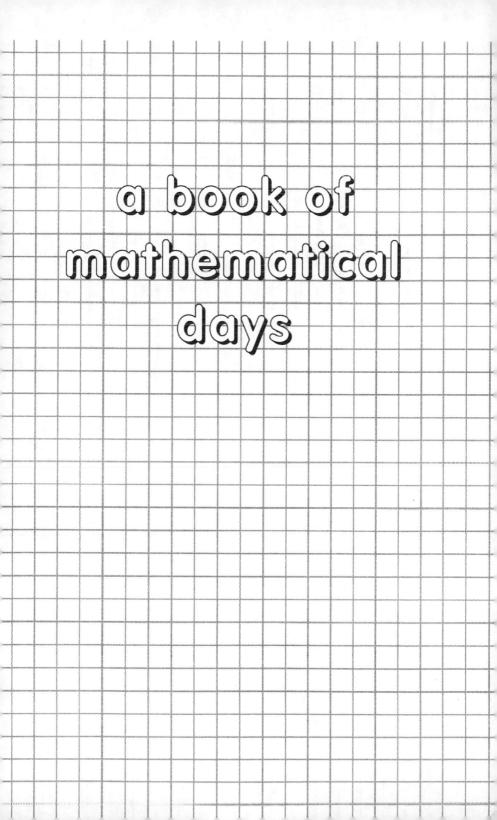

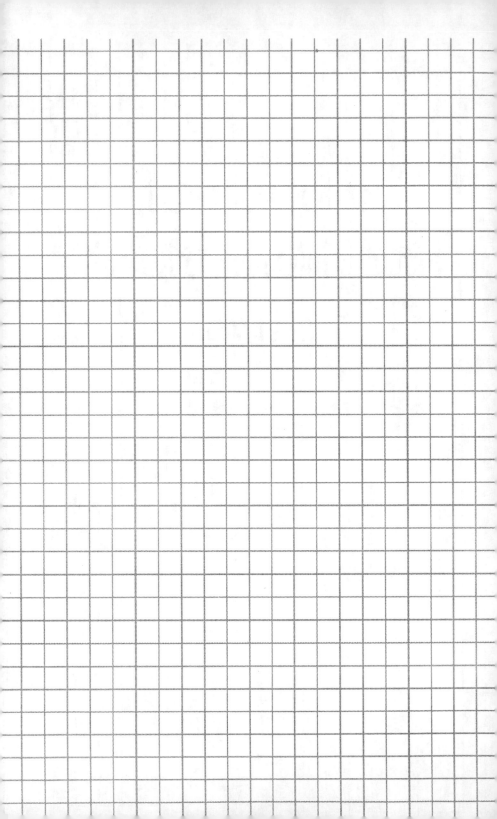

January 1 ●

S M T W T F S

January dates are in the
Mayan number system

*"...zero.... is only forced on us by the
needs of cultivated modes of thought.*
—**Alfred North Whitehead**

HISTORICAL/CURRENT NOTE:
It's *sunya* in Hindu, *cifr* in Arabic
and *zero* in English. The term
cipher refers to an empty
column of beads in an abacus.
Although the Babylonians had
symbols for a zero placeholder
in their numbers, as did the
Mayans, it was the Hindu-Arabic
system that first worked with
the concept of *zero things.*

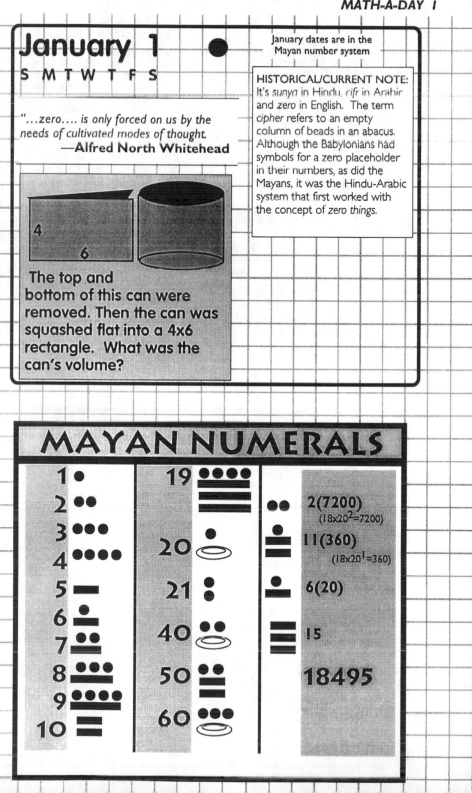

4
6

**The top and
bottom of this can were
removed. Then the can was
squashed flat into a 4x6
rectangle. What was the
can's volume?**

MAYAN NUMERALS

1 •
2 ••
3 •••
4 ••••
5 ▬
6 ▬ with •
7 ▬ with ••
8 ▬ with •••
9 ▬ with ••••
10 ▬▬

19 ••••
20 • ⬭
21 •• (dots)
40 •• ⬭
50 ▬▬ ••
60 ••• ⬭

•• 2(7200)
   $(18 \times 20^2 = 7200)$

▬ with • 11(360)
   $(18 \times 20^1 = 360)$

▬ with • 6(20)

▬▬▬ 15

18495

January 2 ●●

S M T W T F S

Nature uses as little as possible of anything. **—Johannes Kepler**

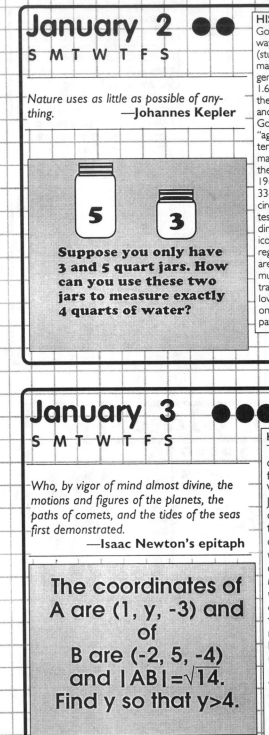

5

3

Suppose you only have 3 and 5 quart jars. How can you use these two jars to measure exactly 4 quarts of water?

HISTORICAL/CURRENT NOTE:
Golfball, you've come a long a long way! You've evolved from a "feathery" (stuffed with feathers) to a "gutty" made from the latex of the Palaquium genus to today's standard 1.62 ounce, 1.68" diameter ball of rubber or synthetic material. The patterns of bumps and indentations evolved by chance. Golfers noticed that "damaged" or "aged" balls with bumps and dents often improved their performance. Enter mathematics and computer analysis of the aerodynamics of the golfball. In the 1960s the most popular pattern had 336 dimpled octahedrons in parallel circles around the ball. After additional testing and analysis of several hundred dimpled configurations and shapes, the icosahedron pattern with 20 triangular regions was developed. Today, there are multi-shaped and sized dimples in multiple patterns. Some give a high trajectory and short roll while others a low trajectory and long roll. The USGA only requires that the golfball's dimple pattern be spherically symmetric.

January 3 ●●●

S M T W T F S

Who, by vigor of mind almost divine, the motions and figures of the planets, the paths of comets, and the tides of the seas first demonstrated.
—Isaac Newton's epitaph

The coordinates of A are (1, y, -3) and of B are (-2, 5, -4) and |AB|=√14. Find y so that y>4.

HISTORICAL/CURRENT NOTE:
The proof of Fermat's Last Theorem was a major mathematical feat of the 20th century. Andrew Wiles first presented his work in June 1993. The great Karl Gauss, on the other hand, had refused to tackle Fermat's Last Theorem, calling it *"an isolated proposition with very little interest for me, because I could easily lay down a multitude of such propositions, which one could neither prove nor dispose of."* There is no question that Gauss was a phenomenal mathematician, but it would have been nice if he had backed up his reply with a few such propositions. It is such propositions that spur new ideas and inventiveness, just as Fermat's Last Theorem has done for centuries.

January 4 ●●●
S M T W T F S

Mathematics consists of proving the most obvious thing in the least obvious way. —**George Polya**

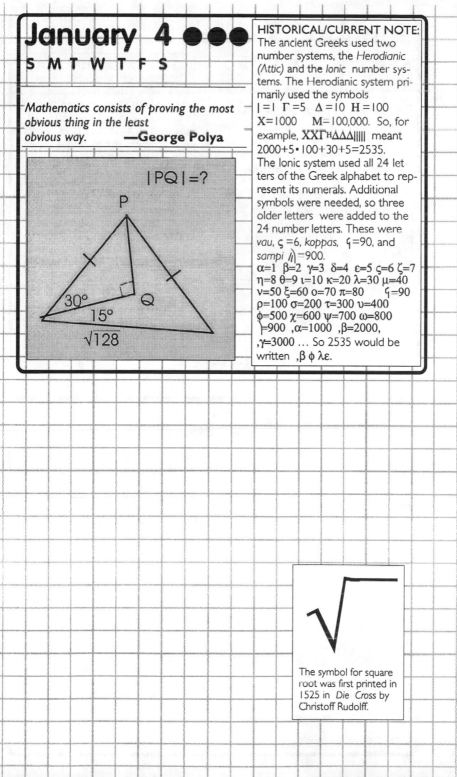

$|PQ|=?$

P

30°

15°

Q

√128

HISTORICAL/CURRENT NOTE:
The ancient Greeks used two number systems, the *Herodianic* (*Attic*) and the *Ionic* number systems. The Herodianic system primarily used the symbols
| =1 Γ =5 Δ =10 H =100
X=1000   M=100,000. So, for example, XXΓᴴΔΔΔ||||| meant 2000+5•100+30+5=2535.
The Ionic system used all 24 letters of the Greek alphabet to represent its numerals. Additional symbols were needed, so three older letters were added to the 24 number letters. These were *vau*, ς =6, *koppas*, ϙ=90, and *sampi* ⋀ =900.
α=1 β=2 γ=3 δ=4 ε=5 ς=6 ζ=7
η=8 θ=9 ι=10 κ=20 λ=30 μ=40
ν=50 ξ=60 ο=70 π=80   ϙ=90
ρ=100 σ=200 τ=300 υ=400
φ=500 χ=600 ψ=700 ω=800
⟋=900 ,α=1000 ,β=2000,
,γ=3000 ... So 2535 would be written ,β φ λε.

The symbol for square root was first printed in 1525 in *Die Cross* by Christoff Rudolff.

January 5

S M T W T F S

...he seemed to approach the grave as an hyperbolic curve approaches a line, less directly as he got nearer, till it was doubtful if he would ever reach it at all.
—Thomas Hardy

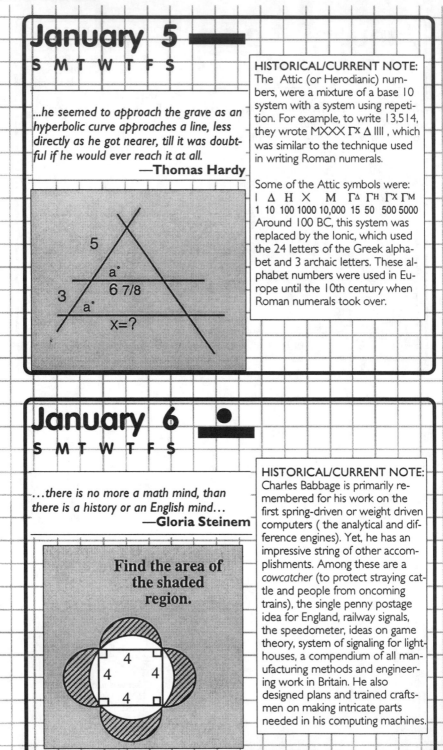

5

a°

6 7/8

3

a°

x=?

HISTORICAL/CURRENT NOTE:
The Attic (or Herodianic) numbers, were a mixture of a base 10 system with a system using repetition. For example, to write 13,514, they wrote MXXX Γˣ Δ IIII , which was similar to the technique used in writing Roman numerals.

Some of the Attic symbols were:
I Δ H X M Γᐃ Γᴴ Γˣ Γᴹ
1 10 100 1000 10,000 15 50 500 5000
Around 100 BC, this system was replaced by the Ionic, which used the 24 letters of the Greek alphabet and 3 archaic letters. These alphabet numbers were used in Europe until the 10th century when Roman numerals took over.

January 6

S M T W T F S

...there is no more a math mind, than there is a history or an English mind...
—Gloria Steinem

Find the area of the shaded region.

4

4   4

4

HISTORICAL/CURRENT NOTE:
Charles Babbage is primarily remembered for his work on the first spring-driven or weight driven computers (the analytical and difference engines). Yet, he has an impressive string of other accomplishments. Among these are a *cowcatcher* (to protect straying cattle and people from oncoming trains), the single penny postage idea for England, railway signals, the speedometer, ideas on game theory, system of signaling for lighthouses, a compendium of all manufacturing methods and engineering work in Britain. He also designed plans and trained craftsmen on making intricate parts needed in his computing machines.

January 7

S M T W T F S

●●
▬▬

If we can imagine a consciousness great enough to know the exact locations and velocities of all the objects in the universe at the present instant, as well as all forces, then there could be no secrets from this consciousness. It could calculate anything about the past or future from the laws of cause and effect.

—**Pierre-Simon de Laplace**

$$\sum_{n=1}^{21} 1 = ?$$

HISTORICAL/CURRENT NOTE: Imagine a book on astronomy chock full of important mathematical ideas. That's what *Bhrama Sphuta Siddhanta* was. Written by astronomer Bhramagupta (598-c.665), it included chapters on both zero and negative numbers, arithmetic rules for zero and positive and negative numbers. In addition, it pointed out that the quadratic equation has two solutions, which was not known to European mathematicians until a millennium after his death.

January 8

S M T W T F S

●●●
▬▬

If we do not expect the unexpected, we will never find it.

—**Heraclitus**

$$32 = ?$$
eight   tcn

HISTORICAL/CURRENT NOTE: Can Mozart improve your mathematical thinking? A University of California at Irvine study showed that students listening to 10 minutes of Mozart's *Sonata for Two Pianos in D Major* (K. 448) raised their IQ scores on spatial temporal reasoning (skills related to mathematics) tests. How does one explain that? It is believed that Mozart strengthens neural connections that underlie mathematical thinking. In addition, some researchers found this sonata also reduced seizures in epileptics and improved spatial temporal reasoning in Alzheimer patients.

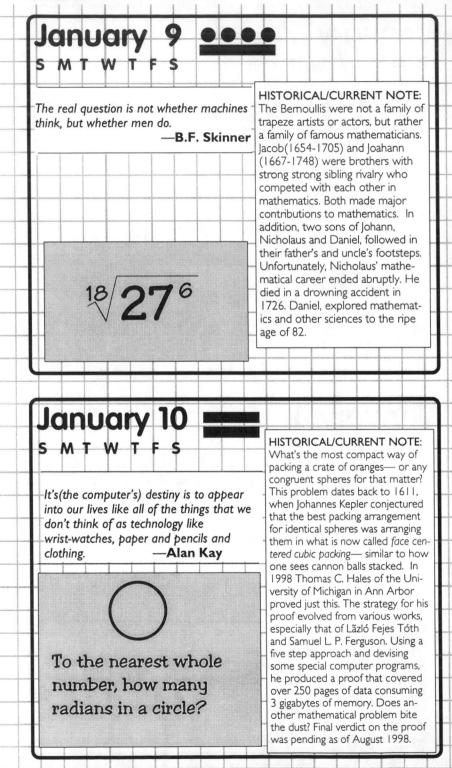

January 9 ●●●●

S M T W T F S

The real question is not whether machines think, but whether men do.

—B.F. Skinner

$$\sqrt[18]{27^6}$$

HISTORICAL/CURRENT NOTE:
The Bernoullis were not a family of trapeze artists or actors, but rather a family of famous mathematicians. Jacob(1654-1705) and Joahann (1667-1748) were brothers with strong strong sibling rivalry who competed with each other in mathematics. Both made major contributions to mathematics. In addition, two sons of Johann, Nicholaus and Daniel, followed in their father's and uncle's footsteps. Unfortunately, Nicholaus' mathematical career ended abruptly. He died in a drowning accident in 1726. Daniel, explored mathematics and other sciences to the ripe age of 82.

January 10 ▬▬

S M T W T F S

It's(the computer's) destiny is to appear into our lives like all of the things that we don't think of as technology like wrist-watches, paper and pencils and clothing. **—Alan Kay**

To the nearest whole number, how many radians in a circle?

HISTORICAL/CURRENT NOTE:
What's the most compact way of packing a crate of oranges— or any congruent spheres for that matter? This problem dates back to 1611, when Johannes Kepler conjectured that the best packing arrangement for identical spheres was arranging them in what is now called *face centered cubic packing*— similar to how one sees cannon balls stacked. In 1998 Thomas C. Hales of the University of Michigan in Ann Arbor proved just this. The strategy for his proof evolved from various works, especially that of Lãzló Fejes Tóth and Samuel L. P. Ferguson. Using a five step approach and devising some special computer programs, he produced a proof that covered over 250 pages of data consuming 3 gigabytes of memory. Does another mathematical problem bite the dust? Final verdict on the proof was pending as of August 1998.

January 11
S M T W T F S

Many who have had an opportunity of knowing any more about mathematics confuse it with arithmetic, and consider it an arid science. In reality, however, it is a science which requires a great amount of imagination. —**Sonya Kovalevsky**

$$\left[(3.62) \div (.079)\right]^{0}$$

HISTORICAL/CURRENT NOTE:
What's so special about Archimedes (c.287-212 BC) of Syracuse and his work? His work dealt with *original* ideas and proofs. The mathematics of these were not extensions of works of others, but found by Archimedes. The breadth of his discoveries is very impressive, and much remains in the original Greek text rather than only in translated forms. Among these books are *Measurements Of The Circle, The Sand Reckoner, Spirals, Conoids & Spheroids, Sphere & Cylinder, Floating bodies, The quadrature of the parabola, The equilibrium of plane figures.*

January 12
S M T W T F S

I tell them that if they will occupy themselves with the study of mathematics they will find in it the best remedy against the lusts of the flesh. —**Thomas Mann**

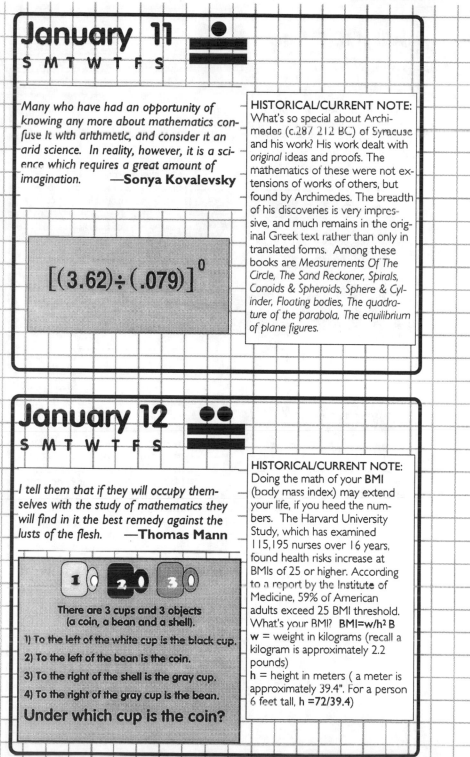

There are 3 cups and 3 objects
(a coin, a bean and a shell).

1) To the left of the white cup is the black cup.

2) To the left of the bean is the coin.

3) To the right of the shell is the gray cup.

4) To the right of the gray cup is the bean.

Under which cup is the coin?

HISTORICAL/CURRENT NOTE:
Doing the math of your **BMI** (body mass index) may extend your life, if you heed the numbers. The Harvard University Study, which has examined 115,195 nurses over 16 years, found health risks increase at BMIs of 25 or higher. According to a report by the Institute of Medicine, 59% of American adults exceed 25 BMI threshold. What's your BMI? $BMI = w/h^2$ B
w = weight in kilograms (recall a kilogram is approximately 2.2 pounds)
h = height in meters (a meter is approximately 39.4"). For a person 6 feet tall, $h = 72/39.4$)

January 13
S M T W T F S

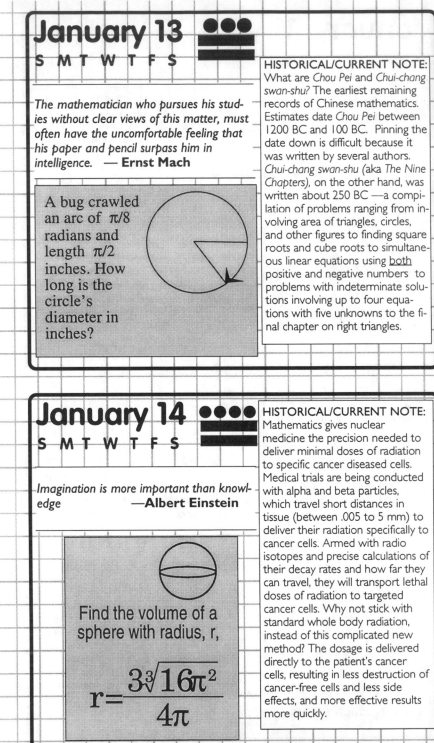

The mathematician who pursues his studies without clear views of this matter, must often have the uncomfortable feeling that his paper and pencil surpass him in intelligence. — **Ernst Mach**

A bug crawled an arc of $\pi/8$ radians and length $\pi/2$ inches. How long is the circle's diameter in inches?

HISTORICAL/CURRENT NOTE:
What are *Chou Pei* and *Chui-chang swan-shu?* The earliest remaining records of Chinese mathematics. Estimates date *Chou Pei* between 1200 BC and 100 BC. Pinning the date down is difficult because it was written by several authors. *Chui-chang swan-shu* (aka *The Nine Chapters*), on the other hand, was written about 250 BC —a compilation of problems ranging from involving area of triangles, circles, and other figures to finding square roots and cube roots to simultaneous linear equations using <u>both</u> positive and negative numbers to problems with indeterminate solutions involving up to four equations with five unknowns to the final chapter on right triangles.

January 14
S M T W T F S

Imagination is more important than knowledge —**Albert Einstein**

Find the volume of a sphere with radius, r,

$$r=\frac{3\sqrt[3]{16\pi^2}}{4\pi}$$

HISTORICAL/CURRENT NOTE:
Mathematics gives nuclear medicine the precision needed to deliver minimal doses of radiation to specific cancer diseased cells. Medical trials are being conducted with alpha and beta particles, which travel short distances in tissue (between .005 to 5 mm) to deliver their radiation specifically to cancer cells. Armed with radio isotopes and precise calculations of their decay rates and how far they can travel, they will transport lethal doses of radiation to targeted cancer cells. Why not stick with standard whole body radiation, instead of this complicated new method? The dosage is delivered directly to the patient's cancer cells, resulting in less destruction of cancer-free cells and less side effects, and more effective results more quickly.

January 15
S M T W T F S

The primary question was not what do we know, but how do we know it.
—Aristotle

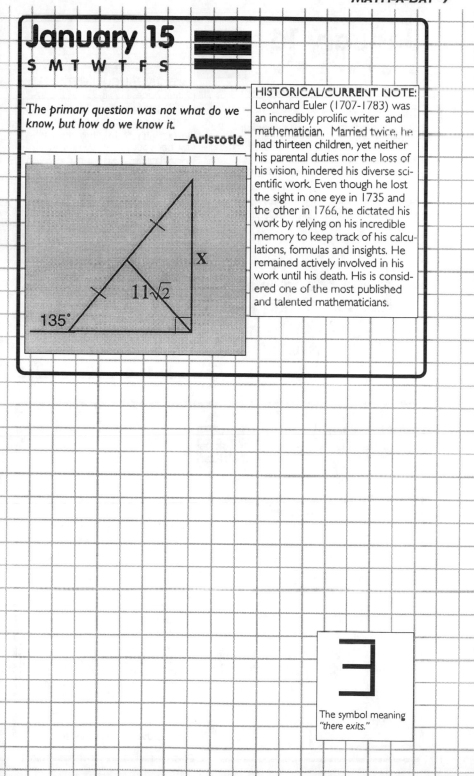

HISTORICAL/CURRENT NOTE:
Leonhard Euler (1707-1783) was an incredibly prolific writer and mathematician. Married twice, he had thirteen children, yet neither his parental duties nor the loss of his vision, hindered his diverse scientific work. Even though he lost the sight in one eye in 1735 and the other in 1766, he dictated his work by relying on his incredible memory to keep track of his calculations, formulas and insights. He remained actively involved in his work until his death. His is considered one of the most published and talented mathematicians.

The symbol meaning
"there exits."

January 16

S M T W T F S

This is the monstrosity of love, that the will is infinite, and the execution confined; that the desire is boundless, and the act a slave to limit.
— **Shakespeare**
Troilus & Cressida

If a regular polygon has exterior angles of measure 18 degrees each, how many sides does it have?

HISTORICAL/CURRENT NOTE:
Maria Agnesi(1718-99), at the age of 20, began writing a book on mathematical ideas which she planned to use as a teaching tool for her younger brothers. Ten years later her lessons had evolved into a two volume book which explained the mathematical ideas on calculus which had surfaced on the European continent and England. When published in 1748, *Analytical Institutions* made quite an impression on the mathematical world. It was the first clear comprehensive textbook on integral and differential calculus which drew on the works of Leibniz, Newton and others, and it was *written by a woman.*

January 17

S M T W T F S

Science is the great antidote to the poison of enthusiasm and superstition.
— **Adam Smith**

$$$$$$$$$$

$20 is deposited for 6 years in an account that yields 5% interest compounded daily. No money is withdrawn. Determine to the nearest dollar how much will be accumulated after six years?

HISTORICAL/CURRENT NOTE:
Did the universe give birth to it-self? Speculations about the ori-gins of the universe have spawned many ideas and theo-ries. Research physicists J. Richard Gott III and Li-Xin Li of Princeton found that quantum effects do not prevent time loops—where traveling back in time won't change what comes later. Li and Gott theorize that the inflation (a rapidly expanding universe) after a big bang could give birth to a baby universe. Then, time loops might allow one of these to be-come the original universe.

January 18
S M T W T F S

A completely unfree society—that is, one proceeding in everything by strict rules of "conformity"—will, in its behavior, be either inconsistent or incomplete, unable to solve certain problems, perhaps of vital importance. —**Kurt Gödel**

$$-\frac{1}{2}x + 0.6(x+5) = 4.4$$

HISTORICAL/CURRENT NOTE: Establishing uniform standard for weights and measures dates back to ancient times. The Egyptians constructed metal bars to correspond to the *Royal cubit* (20.59") and the *short cubit* (17.72"). One of the first major contributions to standardization of units was the Roman law requiring the use of Roman units. The Romans used a base twelve system for measurement, which divided the *foot* and the *pound* into 12 *unciae* (parts). The unit *libra* was used for the weight (hence the abbreviation for the English pound, lb.). In addition, five feet equaled a *pace*, and a *mile* was 1,000 paces. Today, standards for weights and measures are no longer governed by empires or merchants, but by scientific needs.

January 19
S M T W T F S

No one shall expel us from the paradise which Cantor has created us. (Referring to transfinite numbers.) —**David Hilbert**

A car passed me going 10mph faster than my car. How many feet ahead of me will it be in 5 minutes?

HISTORICAL/CURRENT NOTE: In 1974 the Hungarian architect Erno Rubik invented the incredibly popular *Rubik's Cube*, a cube made up of 27 mini cubes which rotate along all three of its axes. When it hit the market in 1975, math enthusiasts immediately began dissecting the mathematics behind it. There are $8.855801027 \times 10^{22}$ possible arrangements with only 2,048 possible solutions.

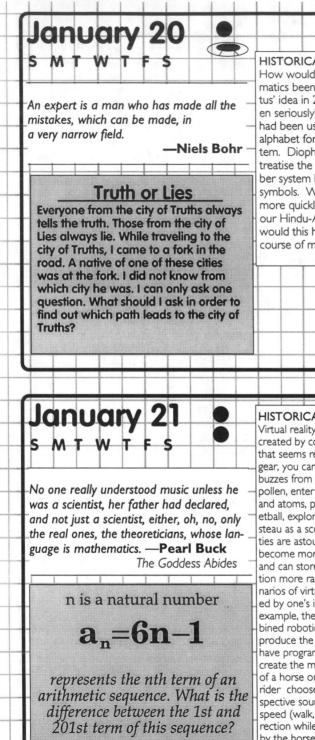

January 20

S M T W T F S

An expert is a man who has made all the mistakes, which can be made, in a very narrow field.

—Niels Bohr

Truth or Lies

Everyone from the city of Truths always tells the truth. Those from the city of Lies always lie. While traveling to the city of Truths, I came to a fork in the road. A native of one of these cities was at the fork. I did not know from which city he was. I can only ask one question. What should I ask in order to find out which path leads to the city of Truths?

HISTORICAL/CURRENT NOTE:
How would the history of mathematics been affected if Diophantus' idea in 250 AD had been taken seriously? The ancient Greeks had been using the letters of their alphabet for their numbering system. Diophantus proposed in a treatise the advantages of a number system based on real number symbols. Would these have led more quickly to the evolution of our Hindu-Arabic numerals? How would this have changed the course of mathematical thinking?

January 21

S M T W T F S

No one really understood music unless he was a scientist, her father had declared, and not just a scientist, either, oh, no, only the real ones, the theoreticians, whose language is mathematics. **—Pearl Buck**
The Goddess Abides

n is a natural number

$$a_n = 6n - 1$$

represents the nth term of an arithmetic sequence. What is the difference between the 1st and 201st term of this sequence?

HISTORICAL/CURRENT NOTE:
Virtual reality is the imaginary world created by computer technology that seems real. Donning the proper gear, you can experience how a bee buzzes from flower to flower for pollen, enter the world of molecules and atoms, play a lively game of racketball, explore the world of Cousteau as a scuba diver. The possibilities are astounding. As computers become more and more powerful and can store and retrieve information more rapidly, the possible scenarios of virtual reality are only limited by one's imagination. For example, the Japanese have combined robotics and virtual reality to produce the virtual horse. They have programmed a computer to re-create the motion and movements of a horse on a robotic horse. The rider chooses the scenery with respective sound effects for the ride, speed (walk, trot or canter) and direction while riding— all controlled by the horse's reins.

January 22

S M T W T F S

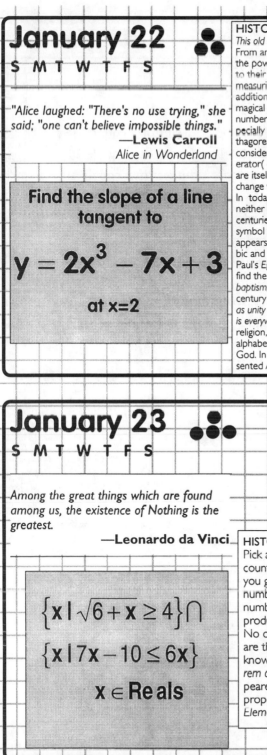

"Alice laughed: "There's no use trying," she said; "one can't believe impossible things."
—**Lewis Carroll**
Alice in Wonderland

Find the slope of a line tangent to

$$y = 2x^3 - 7x + 3$$

at x=2

HISTORICAL/CURRENT NOTE:
This old man he played one. He played…
From ancient times people believed that the powers of numbers was not confined to their use in solving problems, counting, measuring, and describing things; but, in addition, numbers possessed mystical and magical powers. The mystique of the number 1 dates back to ancient times, especially grounded in the beliefs of the Pythagoreans. The ancient Greeks did not consider 1 a number, but rather the generator(or mother) of numbers. Its factors are itself. A number's value does not change when multiplied or divided by 1. In today's mathematical definitions, 1 is neither prime nor composite. Over the centuries the number 1 has become a symbol for the divine. Such a connection appears in Indian, Hebrew, Chinese, Arabic and Egyptian religious writings. In St. Paul's *Epistle to the Ephesians* (4:5), we find the words *"One lord, one faith, one baptism, one God, father of all."* The 17th century mystic Angelus Silesius said, *"Just as unity is every number, thus God the one is everywhere in everything."* In the Muslim religion, 1 is the first letter of the Arabic alphabet, *alif,* and the first letter in *Allah,* God. In early Chinese religion, 1 represented *All, the Perfect, the Absolute.*

January 23

S M T W T F S

Among the great things which are found among us, the existence of Nothing is the greatest.
—**Leonardo da Vinci**

$$\{x \mid \sqrt{6+x} \geq 4\} \cap$$

$$\{x \mid 7x - 10 \leq 6x\}$$

$$x \in \mathbf{Reals}$$

HISTORICAL/CURRENT NOTE:
Pick a counting number, any counting number. Factor it until you get it down to just prime numbers. The string of prime numbers which, when multiplied, produce that number are unique. No other number's prime factors are the same. Today, this is known as *The Fundamental Theorem of Arithmetic.* It first appeared over 2000 years ago as a proposition in book IX of Euclid's *Elements.*

number names and their values

a million equals 10^6 *a billion* equals 10^9

a trillion equals 10^{12}

a quadrillion equals 10^{15} *a quintillion* equals 10^{18}

a sextillion equals 10^{21} *a septillion* equals 10^{24}

an octillion equals 10^{27} *a nonillion* equals 10^{30}

a decillion equals 10^{33} *an undecillion* equals 10^{36}

a duodecillion equals 10^{39} *a tredecillion* equals 10^{42}

a quattuordecillion equals 10^{45}

a quindecillion equals 10^{48}

a sexdecillion equals 10^{51}

a septendecillion equals 10^{54}

a octodecillion equals 10^{57}

a novemdecillion equals 10^{60}

a vigintillion equals 10^{63}

January 24

S M T W T F S

Mathematics ...is as limitless as that space which it finds too narrow for its aspirations; its possiblilities are infinite as the worlds which are forever crowding in and multiplying upon the astronomer's gaze.
—**James J. Sylvester**

$$x + 2y = 19$$

$$x = y + 2z$$

$$3z = 2x - 7$$

find x, y, & z

HISTORICAL/CURRENT NOTE: Where do some of the names of numbers come from? The words *million* and *billion* originated in the late Middle Ages. The ancient Greeks used *myrias* to mean 10,000. In the metric system the prefixes *kilo-, hecto-, deka-* are Greek prefixes while *deci-, centi-,* and *milli-* are Latin. *Trillion* originates from Greek word *teras* meaning monster. The prefix *giga-* (billion), as in gigabyte, comes from the word *gigas* meaning giant; *mega-* (million) comes from *megas* meaning great, *micro-* from *mikros* (millionth) meaning small, while *nano-* (billionth) comes from *nanos* meaning dwarf.

January 25
S M T W T F S

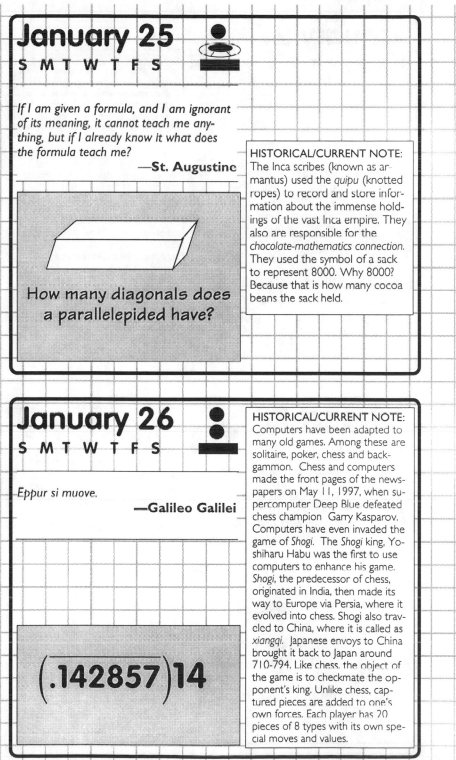

If I am given a formula, and I am ignorant of its meaning, it cannot teach me anything, but if I already know it what does the formula teach me?

—St. Augustine

How many diagonals does a parallelepided have?

HISTORICAL/CURRENT NOTE:
The Inca scribes (known as armantus) used the *quipu* (knotted ropes) to record and store information about the immense holdings of the vast Inca empire. They also are responsible for the *chocolate-mathematics connection.* They used the symbol of a sack to represent 8000. Why 8000? Because that is how many cocoa beans the sack held.

January 26
S M T W T F S

Eppur si muove.

—Galileo Galilei

$$\left(\overline{.142857}\right)14$$

HISTORICAL/CURRENT NOTE:
Computers have been adapted to many old games. Among these are solitaire, poker, chess and backgammon. Chess and computers made the front pages of the newspapers on May 11, 1997, when supercomputer Deep Blue defeated chess champion Garry Kasparov. Computers have even invaded the game of *Shogi*. The *Shogi* king, Yoshiharu Habu was the first to use computers to enhance his game. *Shogi*, the predecessor of chess, originated in India, then made its way to Europe via Persia, where it evolved into chess. Shogi also traveled to China, where it is called as *xiangqi.* Japanese envoys to China brought it back to Japan around 710-794. Like chess, the object of the game is to checkmate the opponent's king. Unlike chess, captured pieces are added to one's own forces. Each player has 20 pieces of 8 types with its own special moves and values.

January 27
S M T W T F S

Chance governs all.

—John Milton

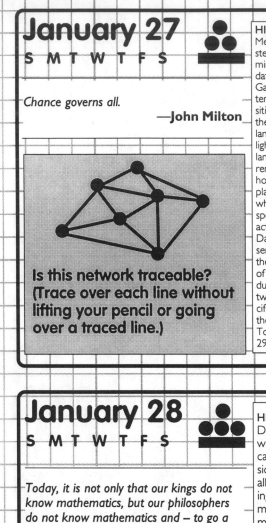

Is this network traceable? (Trace over each line without lifting your pencil or going over a traced line.)

HISTORICAL/CURRENT NOTE:
Mention *the speed of light* and Einstein's universal constant comes to mind. Yet interest in the speed of light dates back centuries before Einstein. Galileo made the first unsuccessful attempt at determining its speed by positioning two lanterns and measuring the delay in response from the second lantern's observer to the first one's light. The distance between the lanterns was double and the delay remained the same. In fact, no matter how far apart the lanterns were placed, the delay was the same—which had nothing to do with the speed of light, but rather the reflex reaction of the experimenters. In 1675, Danish astronomer O. Roemer observed differences in times between the light of the eclipses of the moons of Jupiter, which he determined was due to the different distances between Jupiter and the Earth at a specific time. His calculations determined the speed of light as 240,000 km/sec. Today, its speed has been set at 299,792.458 km/sec.

January 28
S M T W T F S

Today, it is not only that our kings do not know mathematics, but our philosophers do not know mathematics and – to go a step further – our mathematicians do not know mathematics.

— Julius R. Oppenheimer

One solution to a quadratic equation is 1-3i. The coefficients of the equation's terms are rational. What is the constant term when the equation is in standard form?

HISTORICAL/CURRENT NOTE:
During the Dark Ages in Europe women were denied higher education and in some parts even basic reading and writing were not allowed. Consequently, any learning was primarily restricted to the monasteries, and it was here that mathematics could be explored by women. Hroswith, a nun of the Benedictine Abbey in Saxony, distinguished herself as a scholar. She was interested in various fields, including mathematics, and encouraged other nuns to pursue educational work. In fact, her writing showed she understood our solar system many centuries before Newton! She explained that the sun was the center and its gravitational pull held stars in place, like the earth holds its creatures.

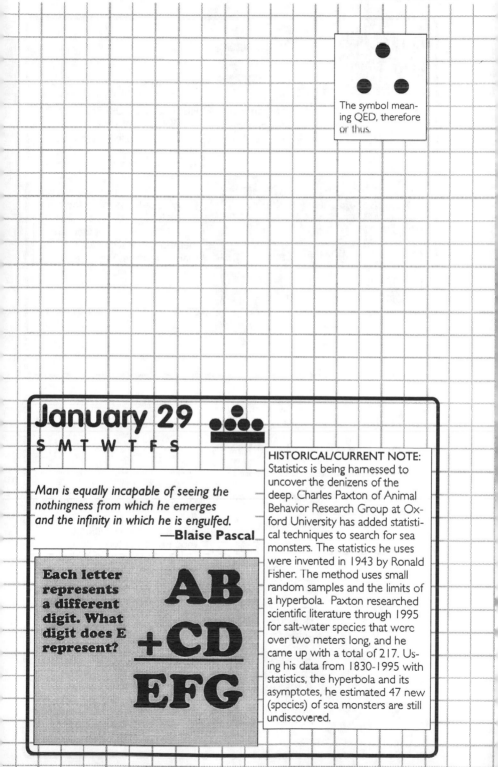

The symbol meaning QED, therefore or thus.

January 29

S M T W T F S

Man is equally incapable of seeing the nothingness from which he emerges and the infinity in which he is engulfed.
—**Blaise Pascal**

Each letter represents a different digit. What digit does E represent?

$$
\begin{array}{r}
AB \\
+\,CD \\
\hline
EFG
\end{array}
$$

HISTORICAL/CURRENT NOTE: Statistics is being harnessed to uncover the denizens of the deep. Charles Paxton of Animal Behavior Research Group at Oxford University has added statistical techniques to search for sea monsters. The statistics he uses were invented in 1943 by Ronald Fisher. The method uses small random samples and the limits of a hyperbola. Paxton researched scientific literature through 1995 for salt-water species that were over two meters long, and he came up with a total of 217. Using his data from 1830-1995 with statistics, the hyperbola and its asymptotes, he estimated 47 new (species) of sea monsters are still undiscovered.

January 30
S M T W T F S

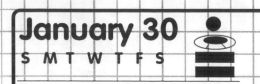

He is unworthy of the name of man who is ignorant of the fact that the diagonal of a square is incommensurable with its side.
—**Plato**

A Lewis Carroll problem
Draw this figure of squares with one continuous line, never crossing any line nor lifting your pencil?

HISTORICAL/CURRENT NOTE:
The Babylonians took π to be 3 1/8, and the Rind Papyrus shows the Egyptians used 256/81. Yet the irrational nature of π was not proven until 1761 by German mathematician Johann Lambert. Carrying out π to more and more decimals places has not lost its appeal . As new computers and methods are discovered, the search continues for patterns hidden in its never ending decimal.

January 31
S M T W T F S

The things of this world cannot be made known without a knowledge of mathematics. —**Roger Bacon**

The base of this pyramid is an equilateral triangle with sides 8 units. The altitude of the pyramid is $\sqrt{3}$. What is its volume?

HISTORICAL/CURRENT NOTE:
Believe it or not, weaving looms have a direct connection to modern computers. The loom was invented by Joseph-Marie Jacquard, and designed to receive its instructions via a program provided by punched cards. Twenty years later, Charles Babbage used such cards in the design of his Analytical Engine to instruct the mathematical operations his machine would perform.

February 11

S M T W T F S

February dates are in the Roman number system

Mathematical genius and artistic genius touch one another.
—Gösta Mittag-Leffler

HISTORICAL/CURRENT NOTE: Mathematics is making its mark on global warming. A joint team from the Center for Climate System Research at the University of Tokyo and The National Institute for Environmental Studies in Japan have developed a numerical model which includes calculations on generating dispersal and chemical reactions of microparticles taking into consideration the effect of wind, rain and seasons. The model correlated closely with actual microparticles' distribution gathered by satellites over a one year period, which presents a more reliable model for making global warming predictions.

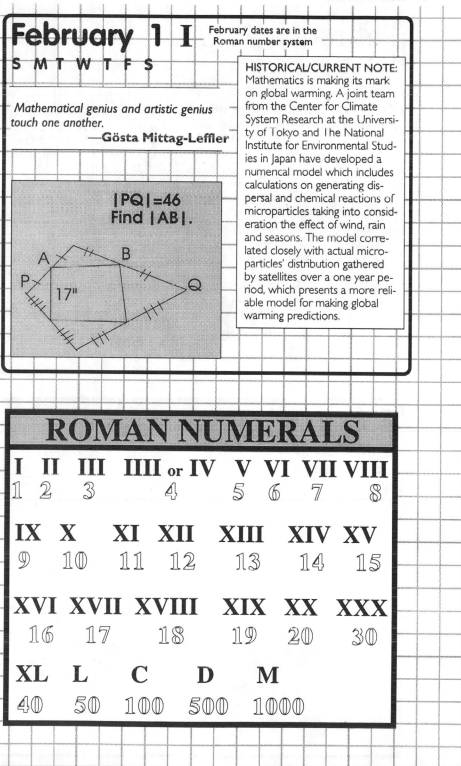

$|PQ| = 46$
Find $|AB|$.

ROMAN NUMERALS

I	II	III	IIII or IV	V	VI	VII	VIII
1	2	3	4	5	6	7	8

IX	X	XI	XII	XIII	XIV	XV
9	10	11	12	13	14	15

XVI	XVII	XVIII	XIX	XX	XXX
16	17	18	19	20	30

XL	L	C	D	M
40	50	100	500	1000

February 2 II
S M T W T F S

If a little knowledge is dangerous, where is the man who has so much as to be out of danger? —**Thomas Huxley**

Two pigeons fly off at the same time and speed. They reach the grain at the same time. How many meters, x, from the foot of the church is the grain?

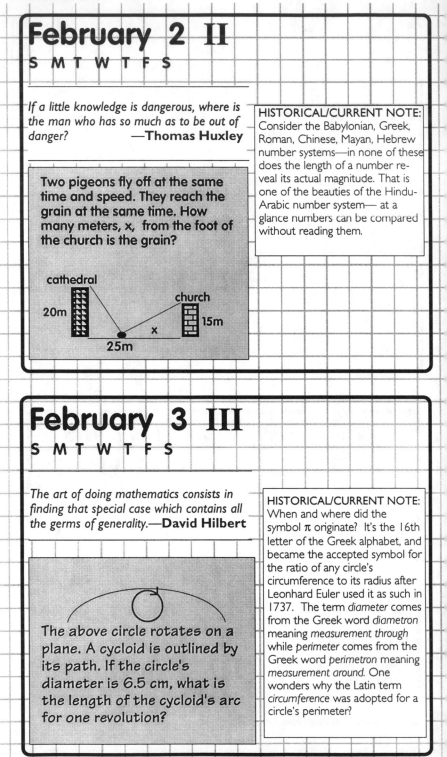

cathedral

20m

church

15m

x

25m

HISTORICAL/CURRENT NOTE: Consider the Babylonian, Greek, Roman, Chinese, Mayan, Hebrew number systems—in none of these does the length of a number reveal its actual magnitude. That is one of the beauties of the Hindu-Arabic number system— at a glance numbers can be compared without reading them.

February 3 III
S M T W T F S

The art of doing mathematics consists in finding that special case which contains all the germs of generality.—**David Hilbert**

The above circle rotates on a plane. A cycloid is outlined by its path. If the circle's diameter is 6.5 cm, what is the length of the cycloid's arc for one revolution?

HISTORICAL/CURRENT NOTE: When and where did the symbol π originate? It's the 16th letter of the Greek alphabet, and became the accepted symbol for the ratio of any circle's circumference to its radius after Leonhard Euler used it as such in 1737. The term *diameter* comes from the Greek word *diametron* meaning *measurement through* while *perimeter* comes from the Greek word *perimetron* meaning *measurement around*. One wonders why the Latin term *circumference* was adopted for a circle's perimeter?

QED

Quod erat demonstrandum is Latin for "which was to be demonstrated". It first appeared in Greek, *οπερ εδει δειξαι*, in the works of Euclid in the 3rd century BC.

February 4 IV

S M T W T F S

Perfect numbers like perfect men are very rare.. **—René Descrates**

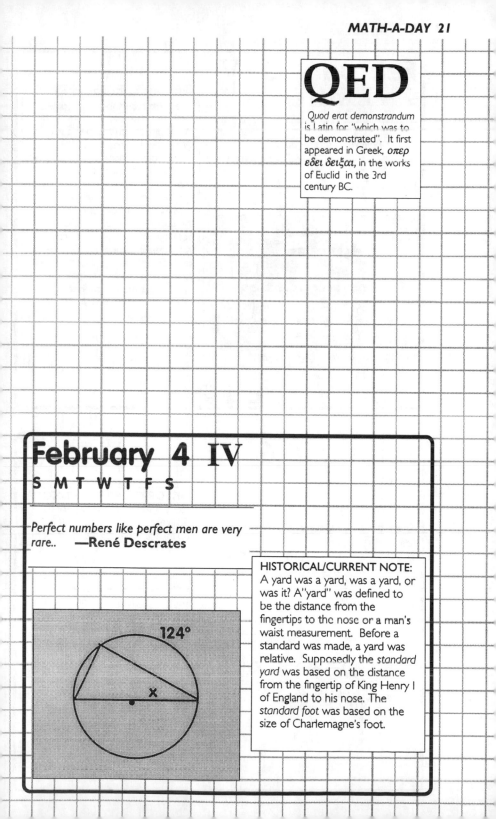

124°

x

HISTORICAL/CURRENT NOTE:
A yard was a yard, was a yard, or was it? A "yard" was defined to be the distance from the fingertips to the nose or a man's waist measurement. Before a standard was made, a yard was relative. Supposedly the *standard yard* was based on the distance from the fingertip of King Henry I of England to his nose. The *standard foot* was based on the size of Charlemagne's foot.

February 5 V

S M T W T F S

I don't believe in mathematics
—Albert Einstein

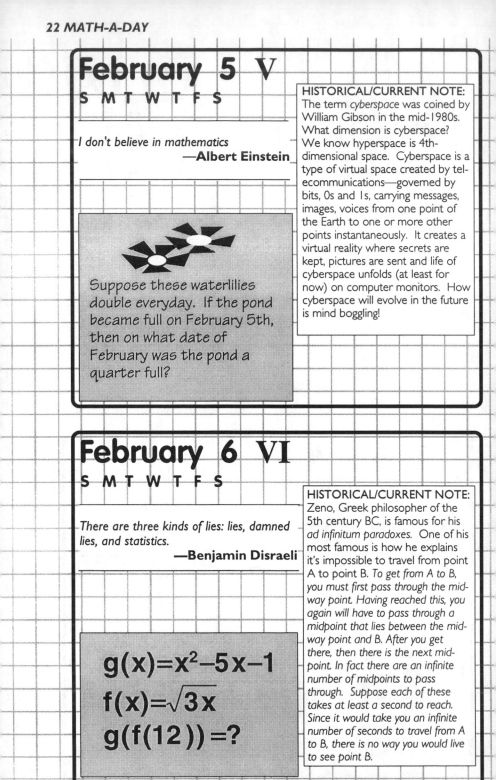

Suppose these waterlilies double everyday. If the pond became full on February 5th, then on what date of February was the pond a quarter full?

HISTORICAL/CURRENT NOTE:
The term *cyberspace* was coined by William Gibson in the mid-1980s. What dimension is cyberspace? We know hyperspace is 4th-dimensional space. Cyberspace is a type of virtual space created by telecommunications—governed by bits, 0s and 1s, carrying messages, images, voices from one point of the Earth to one or more other points instantaneously. It creates a virtual reality where secrets are kept, pictures are sent and life of cyberspace unfolds (at least for now) on computer monitors. How cyberspace will evolve in the future is mind boggling!

February 6 VI

S M T W T F S

There are three kinds of lies: lies, damned lies, and statistics.
—Benjamin Disraeli

$$g(x) = x^2 - 5x - 1$$
$$f(x) = \sqrt{3x}$$
$$g(f(12)) = ?$$

HISTORICAL/CURRENT NOTE:
Zeno, Greek philosopher of the 5th century BC, is famous for his *ad infinitum* paradoxes. One of his most famous is how he explains it's impossible to travel from point A to point B. *To get from A to B, you must first pass through the midway point. Having reached this, you again will have to pass through a midpoint that lies between the midway point and B. After you get there, then there is the next midpoint. In fact there are an infinite number of midpoints to pass through. Suppose each of these takes at least a second to reach. Since it would take you an infinite number of seconds to travel from A to B, there is no way you would live to see point B.*

February 7 VII

S M T W T F S

Poetry is as exact a science as geometry.
—**Gustave Flaubert**

$$30° = ?$$
radians

HISTORICAL/CURRENT NOTE:
Mathematician Gottfried Wilhelm Leibniz is best known as one of the inventors /discoverers of calculus in the late 1600s. In addition, he had a litany of other achievements— he earned a law degree; was a diplomat who tried to reconcile the Protestants and Catholics; he was a philosopher, political writer, advisor to Peter the Great of Russia; he devised the first mechanical calculator that could add, subtract, multiply and divide; he was the first president of the Academy of Sciences in Berlin. Yet, when he died in 1716, his funeral was only attended by his secretary.

February 8 VIII

S M T W T F S

Measure what is measurable, and make measurable what is not so. —**Galileo**

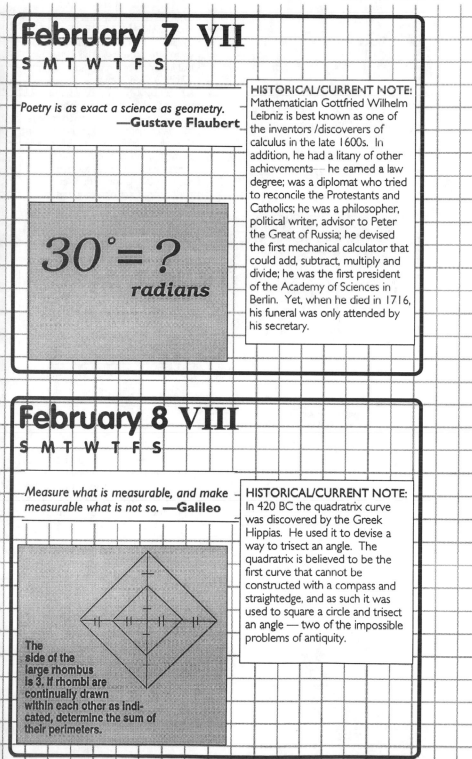

The side of the large rhombus is 3. If rhombi are continually drawn within each other as indicated, determine the sum of their perimeters.

HISTORICAL/CURRENT NOTE:
In 420 BC the quadratrix curve was discovered by the Greek Hippias. He used it to devise a way to trisect an angle. The quadratrix is believed to be the first curve that cannot be constructed with a compass and straightedge, and as such it was used to square a circle and trisect an angle — two of the impossible problems of antiquity.

February 9 IX

S M T W T F S

Mathematics has the completely false reputation of yielding infallible conclusions.. —**Goethe**

$$71.6° \text{ F}$$
$$\text{is}$$
$$\underline{\,?\,} °\text{C}$$

HISTORICAL/CURRENT NOTE: Byzantine monks of the 13th century were into recycling. The oldest original ink copy of Archimedes'(287-121 BC) work, which had been transcribed by a 10th century scribe, was washed off by Byzantine monks of the 13th century in order to recycle the parchment for religious texts. In the Fall of 1998, this surviving copy of mathematical works of Archimedes was auctioned for over $2,000,000. Using ultraviolet light and a computer program, the Archimedean text was dramatically enhanced. The mathematics of the text may reveal new nuances of Archimedes' scientific thinking. The text was sold by a French family who acquired it in the early 1920s. It was purchased by an American, who faces legal battles as to who is the rightful owner—the American buyer or the Greek Orthodox Diocese of Jerusalem.

February 10 X

S M T W T F S

I am interested in mathematics only as a creative art. —**Godfrey Hardy**

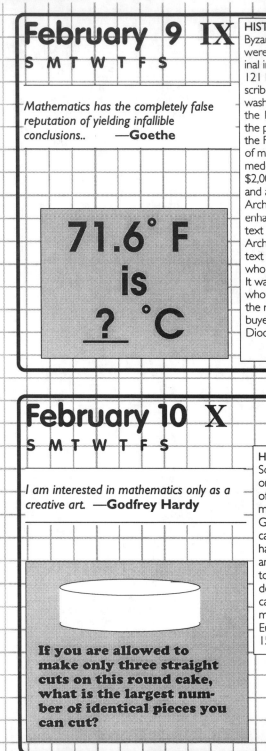

If you are allowed to make only three straight cuts on this round cake, what is the largest number of identical pieces you can cut?

HISTORICAL/CURRENT NOTE: Sometimes stubbornness spites one's own interests. In the case of calculus, it hindered the entire mathematical development of Great Britain. Even though the calculus developed by Leibniz had some superior techniques and notations, the British refused to accept this fact and remained devoted to Newton's form of calculus. As a result, British mathematics remained behind European mathematics for over 150 years!

February 11 XI

S M T W T F S

I admit that mathematical science is a good thing. But excessive devotion to it is a bad thing. **—Aldous Huxley**

$$\frac{2\sqrt{x-4}}{\sqrt{6}} = \sqrt{10}$$

HISTORICAL/CURRENT NOTE:
This old man he played 2, he played… What's the mystique behind the number 2? Consider how many terms there are which express 2— a pair, a couple, twin, double, dual. It's the factor of every even number. Like the number 1, 2 can be considered the generator of even numbers, and is the only even prime number. The duality of many things crosses cultures and adds to the mystique of 2 — here we find the yin and the yang, good and evil, right and wrong, heaven and hell, true or false, Old and New Testaments, binary number system, the expressions two-faced and forked tongue. In Eastern cultures 2 has many negative superstitions. For example, *never do 2 things at once* or *never have 2 related families live in one room.* Jewish tradition says *a man should never pass between 2 women, 2 dogs or 2 pigs,* and *2 men should never let these animals pass by them.* The Latin root *di-* (or *dis-*) found in many words has a divisive or negative connotation— divide, disagree. In the word *doubt* part of the word *double* appears, while doubt in German contains the root for the word *two*.

February 12 XII

S M T W T F S

Descartes commanded the future from his study more than Napoleon from the throne.
— Oliver Wendell Holmes

Money has been stolen. There are four suspects. Only one is telling the truth.
#1 said he didn't steal the money
#2 said #1 was lying
#3 said #2 was lying
#4 said #2 stole the money
Who's telling the truth?

HISTORICAL/CURRENT NOTE:
Is it getting hotter? In addition to revealing the age of a tree, the concentric circles of a tree's trunk also reveal such things as climate fluctuations and droughts. This information, along with glacial ice data and the use of modern instruments and computer technology, led to a recent analysis of the northern hemisphere's temperature by Michael Mann and his colleagues at the University of Massachusetts at Amherst. Their data showed that 1998 was not only the warmest year in the past few centuries, but in the millennium and that the 1990s was the warmest decade. Are you sceptical? Check out their numbers and their charts for yourself.

February 13 XIII

S M T W T F S

...sole end of science is the honor of the human mind, and that under this title a question about numbers is worth as much as a question about the system of the world. —**Carl Jacobi**

$$3+\cfrac{12}{\cfrac{5}{1+\cfrac{2}{1-\frac{1}{2}}}}$$

HISTORICAL/CURRENT NOTE:
The ancient Greeks would not accept the existence of numbers less than zero. In the 3rd century AD, Diophantus rejected "–4" as a solution of an equation. In fact, the first mathematician to systematically use numbers less than zero was Italian Girolamo Cardano. He felt they were fictitious numbers, but he explained why a $-\bullet-$ $=+$. The term "negative" number is of Latin origin meaning "to deny"; perhaps used because the existence of negative had been denied for so many years.

February 14 XIV

S M T W T F S

The so-called Pythagoreans... fancied that the principles of mathematics were the principles of all things. — **Aristotle**

If the difference of the squares of two consecutive natural numbers equals 15, what is the smaller natural number?

HISTORICAL/CURRENT NOTE:
Showing something is impossible is just as important as showing how to do something. The first record of the demonstration of a mathematical impossibility was done by Karl Gauss. As a teenager at the University of Göttingen, he showed how to construct a regular 17-sided polygon using only a straightedge and compass, but more importantly he proved which polygons could not be constructed this way.

π

February 15 XV
S M T W T F S

Now I feel as if I should succeed in doing something in mathematics, although I cannot see why it is so very important... The knowledge doesn't make life any sweeter or happier, does it? **—Helen Keller**

HISTORICAL/CURRENT NOTE:
An electromechanical combination lock on a computer chip? What next? To lock out hackers and cyberthiefs, computer engineers have done just that. Out of a million possible combinations, this lock (each of 6 gears no larger than a period) accepts only one correct combination. If someone tries, only once, to unlock it with the wrong code, the lock disconnects the computer from the network, and can only be reopened by someone who is physically present at the computer. The device was designed by Frank Peter of Sandia National Laboratories in Albuquerque, NM. The device will be commercially available eventually.

Find x.

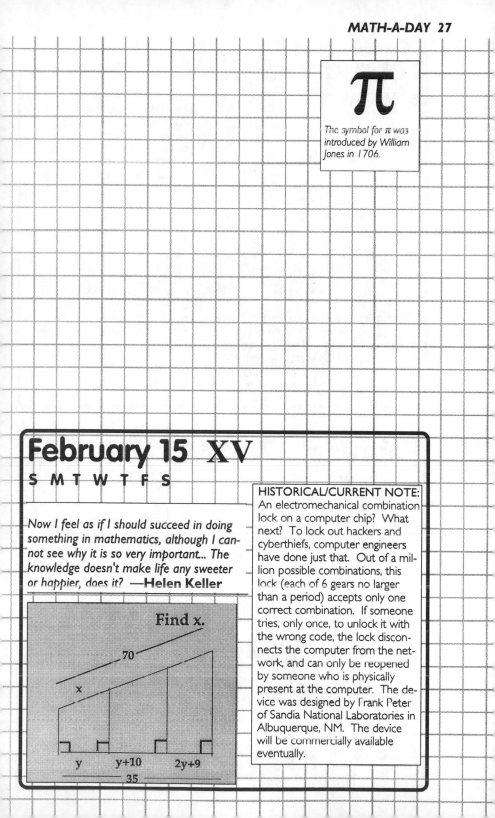

70

x

y   y+10   2y+9

35

February 16 XIV

S M T W T F S

Six is a number perfect in itself, and not because God created the world in six days; rather the contrary is true. God created the world in six days because this number is perfect, and it would remain perfect, even if the work of the six days did not exist.
—St. Augustine

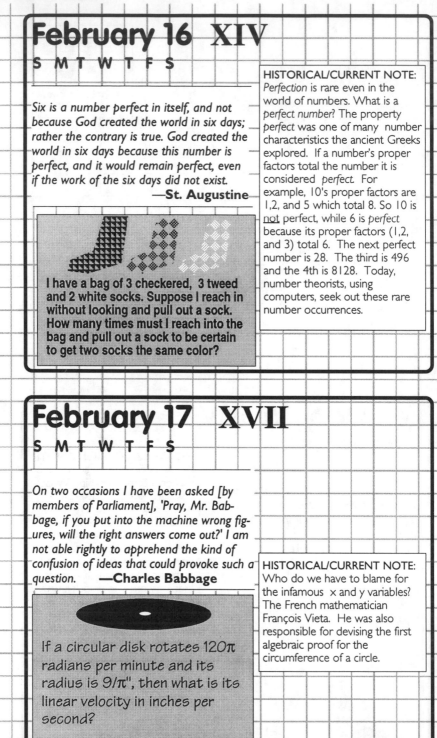

I have a bag of 3 checkered, 3 tweed and 2 white socks. Suppose I reach in without looking and pull out a sock. How many times must I reach into the bag and pull out a sock to be certain to get two socks the same color?

HISTORICAL/CURRENT NOTE:
Perfection is rare even in the world of numbers. What is a *perfect number?* The property *perfect* was one of many number characteristics the ancient Greeks explored. If a number's proper factors total the number it is considered *perfect.* For example, 10's proper factors are 1,2, and 5 which total 8. So 10 is not perfect, while 6 is *perfect* because its proper factors (1,2, and 3) total 6. The next perfect number is 28. The third is 496 and the 4th is 8128. Today, number theorists, using computers, seek out these rare number occurrences.

February 17 XVII

S M T W T F S

On two occasions I have been asked [by members of Parliament], 'Pray, Mr. Babbage, if you put into the machine wrong figures, will the right answers come out?' I am not able rightly to apprehend the kind of confusion of ideas that could provoke such a question. **—Charles Babbage**

If a circular disk rotates 120π radians per minute and its radius is 9/π", then what is its linear velocity in inches per second?

HISTORICAL/CURRENT NOTE:
Who do we have to blame for the infamous x and y variables? The French mathematician François Vieta. He was also responsible for devising the first algebraic proof for the circumference of a circle.

February 18 XVIII

S M T W T F S

I'm sorry to say that the subject I most disliked was mathematics. I have thought about it. I think the reason was that mathematics leaves no room for argument. If you made a mistake, that was all there was to it. **—Malcom X**

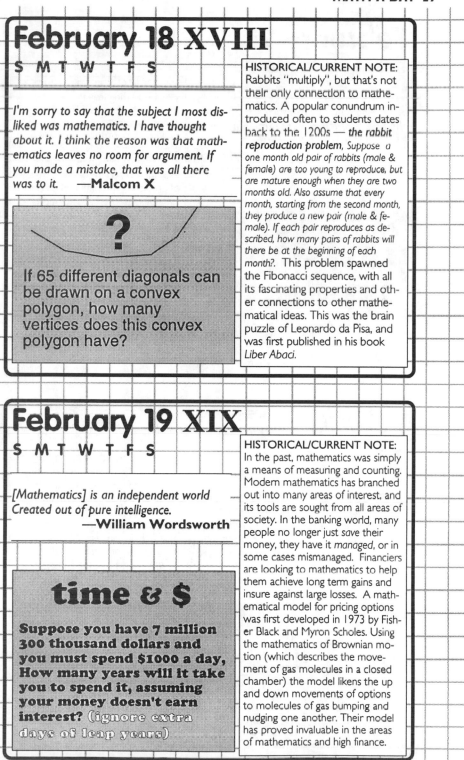

If 65 different diagonals can be drawn on a convex polygon, how many vertices does this convex polygon have?

HISTORICAL/CURRENT NOTE: Rabbits "multiply", but that's not their only connection to mathematics. A popular conundrum introduced often to students dates back to the 1200s — *the rabbit reproduction problem, Suppose a one month old pair of rabbits (male & female) are too young to reproduce, but are mature enough when they are two months old. Also assume that every month, starting from the second month, they produce a new pair (male & female). If each pair reproduces as described, how many pairs of rabbits will there be at the beginning of each month?.* This problem spawned the Fibonacci sequence, with all its fascinating properties and other connections to other mathematical ideas. This was the brain puzzle of Leonardo da Pisa, and was first published in his book *Liber Abaci.*

February 19 XIX

S M T W T F S

[Mathematics] is an independent world Created out of pure intelligence.
—William Wordsworth

time & $

Suppose you have 7 million 300 thousand dollars and you must spend $1000 a day, How many years will it take you to spend it, assuming your money doesn't earn interest? (ignore extra days of leap years)

HISTORICAL/CURRENT NOTE: In the past, mathematics was simply a means of measuring and counting. Modern mathematics has branched out into many areas of interest, and its tools are sought from all areas of society. In the banking world, many people no longer just *save* their money, they have it *managed*, or in some cases mismanaged. Financiers are looking to mathematics to help them achieve long term gains and insure against large losses. A mathematical model for pricing options was first developed in 1973 by Fisher Black and Myron Scholes. Using the mathematics of Brownian motion (which describes the movement of gas molecules in a closed chamber) the model likens the up and down movements of options to molecules of gas bumping and nudging one another. Their model has proved invaluable in the areas of mathematics and high finance.

February 20 XX

S M T W T F S

...An eloquent mathematician must, from the nature of things, ever remain as rare a phenomenon as a talking fish,...
—**J.J. Sylvester**

$$\frac{1}{5} y = 5 \sin 2\theta$$

has amplitude equal to ? .

HISTORICAL/CURRENT NOTE:
What do falling leaves, somersaulting coins falling through water, parachute relief food packages , and other such falling objects have to do with mathematics? Mathematicians are exploring chaos theory as a means of predicting the paths taken by such objects. Perhaps solutions to such problems will be feasible in the future with the ever growing power of supercomputers.

February 21 XXI

S M T W T F S

A man is like a fraction whose numerator is what he is and whose denominator is what he thinks of himself. The larger the denominator the smaller the fraction.
—**Tolstoy**

arithmetic means & ratios

54, x, y, 43, and 21 have arithmetic mean 40.4. The ratio of x and y is 6. Find the values of x and y.

HISTORICAL/CURRENT NOTE:
The Old Testament of the Bible has a number of references to large quantities. Rather than naming a particular number, comparisons are made. For example in Genesis 22:17 a quantity is compared to *the stars of the heaven and the sands;* in Kings 3:8 we find the phrase *"cannot be numbered or counted".* The largest named number appears in 2 Chronicles 14:9 when the term *a thousand thousands* is used. The word a *million* first appears in the late Middle Ages. When a *thousand-thousands* became a commonly used expression by merchants in the late Middle Ages, the term *million* was coined from the Italian word *milione,* meaning *larger than a thousand.*

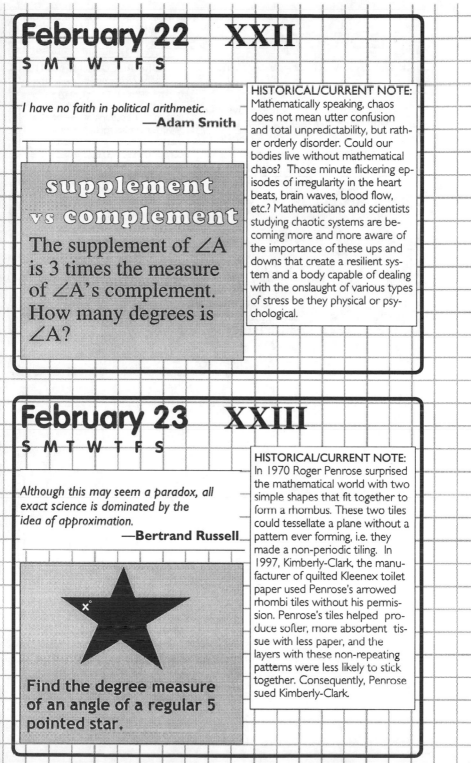

February 22 XXII

S M T W T F S

I have no faith in political arithmetic.
—Adam Smith

supplement vs complement

The supplement of ∠A
is 3 times the measure
of ∠A's complement.
How many degrees is
∠A?

HISTORICAL/CURRENT NOTE:
Mathematically speaking, chaos
does not mean utter confusion
and total unpredictability, but rather orderly disorder. Could our
bodies live without mathematical
chaos? Those minute flickering ep-
isodes of irregularity in the heart
beats, brain waves, blood flow,
etc.? Mathematicians and scientists
studying chaotic systems are becoming more and more aware of
the importance of these ups and
downs that create a resilient system and a body capable of dealing
with the onslaught of various types
of stress be they physical or psy-
chological.

February 23 XXIII

S M T W T F S

*Although this may seem a paradox, all
exact science is dominated by the
idea of approximation.*
—Bertrand Russell

Find the degree measure
of an angle of a regular 5
pointed star.

HISTORICAL/CURRENT NOTE:
In 1970 Roger Penrose surprised
the mathematical world with two
simple shapes that fit together to
form a rhombus. These two tiles
could tessellate a plane without a
pattern ever forming, i.e. they
made a non-periodic tiling. In
1997, Kimberly-Clark, the manu-
facturer of quilted Kleenex toilet
paper used Penrose's arrowed
rhombi tiles without his permis-
sion. Penrose's tiles helped produce softer, more absorbent tissue with less paper, and the
layers with these non-repeating
patterns were less likely to stick
together. Consequently, Penrose
sued Kimberly-Clark.

February 24   XXIV

S M T W T F S

A scientist worthy of his name, above all a mathematician, experiences in his work the same impression as an artist; his pleasure is as great and of the same nature.

—**Henri Poincaré**

It was Monday , and Tom and Jerry were at the same job. When it came time for Tom to go home, Jerry wouldn't let him? Why?

HISTORICAL/CURRENT NOTE:
Encryption technology is one of the hottest areas of study today. It keeps things private that are transmitted electronically — be they government secrets, money matters, medical information, corporate secrets, etc.. An old technique in new packaging is being explored. Recall how a message could be hidden within a message, and only those with the punched out cards could figure out what was the intended secret. That design technique is being explored with electronic messages. The sender parcels out the intended secret in little tagged unscrambled electronic packages which are mixed into other electronic packages. The recipient knows how to eliminate the false packages and hold onto the important bits of information The method is call *chaffing & winnowing*.

$$\leq$$

Pierre Bouguer first introduced the less than or equal & the greater than or equal symbols in 1734. He wrote them as ≦ or ≧.

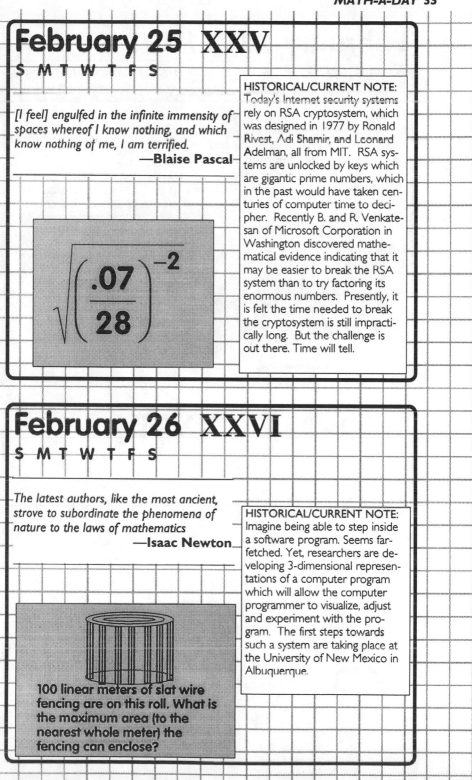

February 25 XXV
S M T W T F S

[I feel] engulfed in the infinite immensity of spaces whereof I know nothing, and which know nothing of me, I am terrified.
—Blaise Pascal

$$\sqrt{\left(\frac{.07}{28}\right)^{-2}}$$

HISTORICAL/CURRENT NOTE:
Today's Internet security systems rely on RSA cryptosystem, which was designed in 1977 by Ronald Rivest, Adi Shamir, and Leonard Adelman, all from MIT. RSA systems are unlocked by keys which are gigantic prime numbers, which in the past would have taken centuries of computer time to decipher. Recently B. and R. Venkatesan of Microsoft Corporation in Washington discovered mathematical evidence indicating that it may be easier to break the RSA system than to try factoring its enormous numbers. Presently, it is felt the time needed to break the cryptosystem is still impractically long. But the challenge is out there. Time will tell.

February 26 XXVI
S M T W T F S

The latest authors, like the most ancient, strove to subordinate the phenomena of nature to the laws of mathematics
—Isaac Newton

HISTORICAL/CURRENT NOTE:
Imagine being able to step inside a software program. Seems farfetched. Yet, researchers are developing 3-dimensional representations of a computer program which will allow the computer programmer to visualize, adjust and experiment with the program. The first steps towards such a system are taking place at the University of New Mexico in Albuquerque.

100 linear meters of slat wire fencing are on this roll. What is the maximum area (to the nearest whole meter) the fencing can enclose?

February 27 XXVII

S M T W T F S

Although the whole of this life were said to be nothing but a dream and the physical world nothing but a phantasm, I should call this dream or phantasm real enough, if, using reason well, we were never deceived by it. —**Gottfried Leibniz**

$$333 = 171$$

base ?   base 10

HISTORICAL/CURRENT NOTE:
What does $\infty - \infty = ?$ or $\infty + \infty = ?$ It's ∞. Consider the infinite set of natural numbers $\{1,2,3,4,\ldots\}$. Take away the infinite set of odd numbers from it, we still have left the infinite set of even numbers. Double the infinite set of natural numbers, we still have an infinite set $\{1,1,2,2,3,3,4,4,\ldots\}$. For thousands of years infinity has never ceased to astound and foil mathematicians. Zeno explored his infinity paradoxes, Galileo was intrigued by infinity. It was Georg Cantor's (1845-1918) phenomenal work in set theory and transfinite numbers that opened up mathematical thought to new ways to think about infinity.

February 28 XXVIII

S M T W T F S

Medicine makes people ill, mathematics make them sad and theology makes them sinful.

—**Martin Luther**

$$\frac{\sum_{n=1}^{?} 3n-10}{25} = 7$$

HISTORICAL/CURRENT NOTE:
Playing cards were designed around 800 A.D. in Hindustan, and appeared in Europe in 1279 in Italian. Later, they made their way to Germany, France and Spain. Their design evolved with a mathematical twist. Any card from a deck of 52 has symmetry, and in fact both vertical and horizontal rotational symmetry. No matter how they are delt the recipient does not have to rotate them to view them right side up.

February 29 XXIX

S M T W T F S

Feb. 29th is only for leap years. These are century years divisible by 400 and all other years divisible by 4.

Logic is the art of going wrong with confidence.. —**Morris Kline**

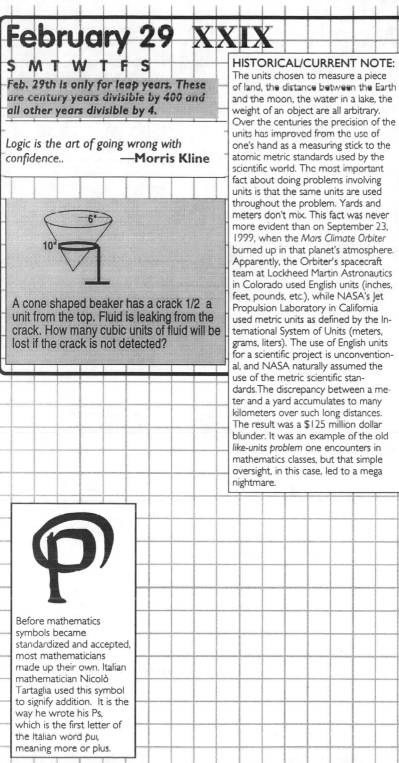

A cone shaped beaker has a crack 1/2 a unit from the top. Fluid is leaking from the crack. How many cubic units of fluid will be lost if the crack is not detected?

HISTORICAL/CURRENT NOTE:

The units chosen to measure a piece of land, the distance between the Earth and the moon, the water in a lake, the weight of an object are all arbitrary. Over the centuries the precision of the units has improved from the use of one's hand as a measuring stick to the atomic metric standards used by the scientific world. The most important fact about doing problems involving units is that the same units are used throughout the problem. Yards and meters don't mix. This fact was never more evident than on September 23, 1999, when the *Mars Climate Orbiter* burned up in that planet's atmosphere. Apparently, the Orbiter's spacecraft team at Lockheed Martin Astronautics in Colorado used English units (inches, feet, pounds, etc.), while NASA's Jet Propulsion Laboratory in California used metric units as defined by the International System of Units (meters, grams, liters). The use of English units for a scientific project is unconventional, and NASA naturally assumed the use of the metric scientific standards. The discrepancy between a meter and a yard accumulates to many kilometers over such long distances. The result was a $125 million dollar blunder. It was an example of the old *like-units problem* one encounters in mathematics classes, but that simple oversight, in this case, led to a mega nightmare.

Before mathematics symbols became standardized and accepted, most mathematicians made up their own. Italian mathematician Nicolò Tartaglia used this symbol to signify addition. It is the way he wrote his Ps, which is the first letter of the Italian word *pui*, meaning more or plus.

March 1

S M T W T F S

He who loves practice without theory is like the sailor who boards ship without a rudder and compass and never knows where he may be cast. —**Leonardo da Vinci**

$$-1-9^{\frac{1}{2}}+8^{\frac{1}{3}}+9^{\frac{1}{2}}$$

January dates are in the Chinese rod numerals

HISTORICAL/CURRENT NOTE: We almost take counting for granted, yet it was this elementary concept that launched the field of mathematics. Today, banks, grocery stores, calculators, computers take care of most of our counting needs. Yet, in the Neolithic period, counting was the highest form of mathematical thinking. Later, people began counting the passage of days, the cycles of the moons, the fruits of their farming or hunting. Mathematical ideas began to surface in crafts and buildings or shelters. Geometric shapes and designs were used to adorn pottery and baskets. Contact with other people necessitated bartering, and the development of various types and denomination of bartering items. With these, the concept of numbers and numbers system were invented. Mathematics began to evolve and connect to our lives.

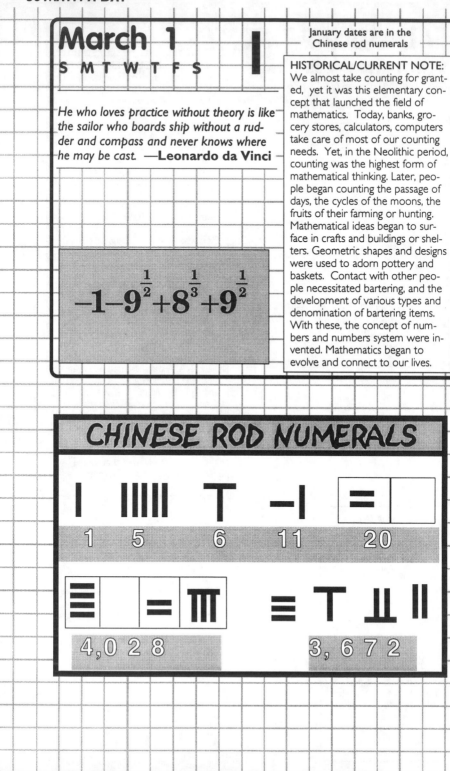

CHINESE ROD NUMERALS

1	5	6	11	20

4,028

3,672

March 2
S M T W T F S

Mathematics is a figment of the imagination. **Theoni Pappas**

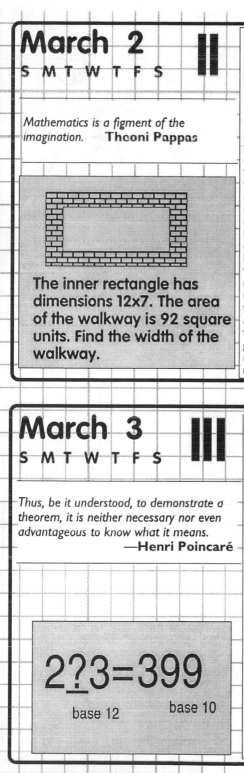

The inner rectangle has dimensions 12x7. The area of the walkway is 92 square units. Find the width of the walkway.

HISTORICAL/CURRENT NOTE:
Suppose someone from the future found a book, and it was the only book with mathematical problems and ideas left from the present. That is exactly what the discovery of the Rhind and Golonishev papyrus resemble — mathematical writing from ancient Egypt. We cannot expect them to represent all the mathematical knowledge of that time, but it does tell us some of the things Egyptians were doing in mathematics. They had a decimal notation system with a unit symbol which did not have place value or a zero symbol. Hieroglyphic and hieratic(cursive) symbols were used to denote various powers of ten. A number was written by using an accumulation of the symbols. These papyruses contained a wealth of problems which used both whole numbers and fractions in their solutions. Among these are problems dealing with basic operations, linear equations, tables, practical geometric problems involving circles, rectangles, triangles and pyramids, and measurements and inventories.

March 3
S M T W T F S

Thus, be it understood, to demonstrate a theorem, it is neither necessary nor even advantageous to know what it means.
—**Henri Poincaré**

$$2?3 = 399$$

base 12   base 10

HISTORICAL/CURRENT NOTE:
This old man he played 3. He played... What's the mystique behind the number 3, the first odd prime number? The first polygon has 3 sides. 3 plays many roles in religion and mythology. Among the terms meaning 3 are trinity, tri- triple triad, trio. In Christianity, there is the Trinity. In Greek mythology, we find the 3 Furies, 3 Graces, 3 Fates, while in other literature we have 3 wishes, the 3 wisemen, 3 strikes and you're out, 3's a charm, Peter's denial of Christ 3 times, Lewis Carroll's phrase "What I tell you 3 times is true" in the *Hunting of the Snark*, *The 3 Little Pigs*, *The Three Little Kittens*, the superstition that things happen in 3s, etc. More of its mystique comes from such ideas as the 3 elements (animal, vegetable, mineral), 3 states (gaseous, liquid, solid), 3-dimensions of space. Among the civilizations that refer to the importance of 3 are the Babylonian, Egyptian, Greek, Chinese, and Indian.

The letter "e" first appeared in print in 1736 in Leonhard Euler's work *Mechanica*.

March 4

S M T W T F S

IIII

Nothing has afforded me so convincing a proof of the unity of the Deity as these purely mental conceptions of numerical and mathematical sciences which have been by slow degrees vouchsafed to man, and are still granted in these later times by the Differential Calculus, now superseded by the Higher Algebra, all of which must have existed in that sublimely omniscient Mind from Eternity. **—Mary Somerville**

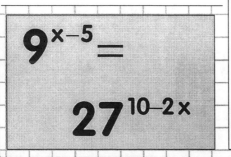

$$9^{x-5} = 27^{10-2x}$$

HISTORICAL/CURRENT NOTE: The origin of prime numbers and methods for finding them dates back to ancient times with the sieve of Eratosthenes (276-194 B.C.), a simple yet ingenious method for sifting out these important numbers. Ever since Euclid proved there were infinitely many primes, the hunt for these numbers has not ceased. Among the famous hunters we find Euclid, Marin Mersenne, Pierre de Fermat, Leonhard Euler, Karl Gauss, and Sophie Germaine. Today, the chase is more frantic than ever. Computers have boosted the size of found primes to enormous sizes. What do you do with a a gigantic prime number once you find it? Set a new record, and show how security codes and ciphers can become obsolete very quickly.

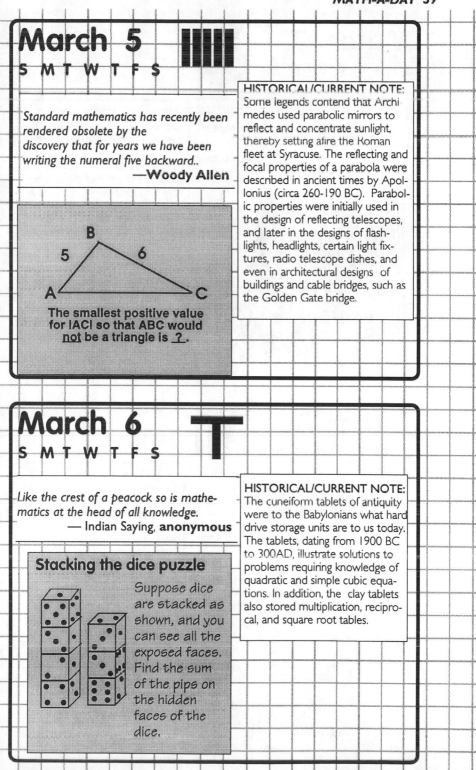

March 5
S M T W T F S

Standard mathematics has recently been rendered obsolete by the discovery that for years we have been writing the numeral five backward..
—Woody Allen

B

5   6

A   C

The smallest positive value for |AC| so that ABC would _not_ be a triangle is _?_.

HISTORICAL/CURRENT NOTE:
Some legends contend that Archimedes used parabolic mirrors to reflect and concentrate sunlight, thereby setting afire the Roman fleet at Syracuse. The reflecting and focal properties of a parabola were described in ancient times by Apollonius (circa 260-190 BC). Parabolic properties were initially used in the design of reflecting telescopes, and later in the designs of flashlights, headlights, certain light fixtures, radio telescope dishes, and even in architectural designs of buildings and cable bridges, such as the Golden Gate bridge.

March 6
S M T W T F S

Like the crest of a peacock so is mathematics at the head of all knowledge.
— Indian Saying, **anonymous**

Stacking the dice puzzle

Suppose dice are stacked as shown, and you can see all the exposed faces. Find the sum of the pips on the hidden faces of the dice.

HISTORICAL/CURRENT NOTE:
The cuneiform tablets of antiquity were to the Babylonians what hard drive storage units are to us today. The tablets, dating from 1900 BC to 300AD, illustrate solutions to problems requiring knowledge of quadratic and simple cubic equations. In addition, the clay tablets also stored multiplication, reciprocal, and square root tables.

March 7

S M T W T F S

π

> It is not once nor twice but times without number that the same ideas make their appearance in the world.
>
> **—Aristotle**

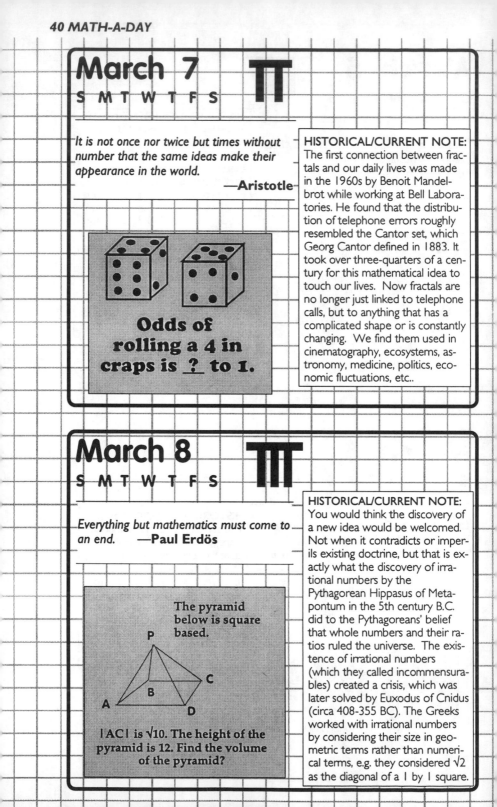

Odds of rolling a 4 in craps is _?_ to 1.

HISTORICAL/CURRENT NOTE: The first connection between fractals and our daily lives was made in the 1960s by Benoit Mandelbrot while working at Bell Laboratories. He found that the distribution of telephone errors roughly resembled the Cantor set, which Georg Cantor defined in 1883. It took over three-quarters of a century for this mathematical idea to touch our lives. Now fractals are no longer just linked to telephone calls, but to anything that has a complicated shape or is constantly changing. We find them used in cinematography, ecosystems, astronomy, medicine, politics, economic fluctuations, etc..

March 8

S M T W T F S

π

> Everything but mathematics must come to an end. **—Paul Erdös**

The pyramid below is square based.

|AC| is √10. The height of the pyramid is 12. Find the volume of the pyramid?

HISTORICAL/CURRENT NOTE: You would think the discovery of a new idea would be welcomed. Not when it contradicts or imperils existing doctrine, but that is exactly what the discovery of irrational numbers by the Pythagorean Hippasus of Metapontum in the 5th century B.C. did to the Pythagoreans' belief that whole numbers and their ratios ruled the universe. The existence of irrational numbers (which they called incommensurables) created a crisis, which was later solved by Euxodus of Cnidus (circa 408-355 BC). The Greeks worked with irrational numbers by considering their size in geometric terms rather than numerical terms, e.g. they considered √2 as the diagonal of a 1 by 1 square.

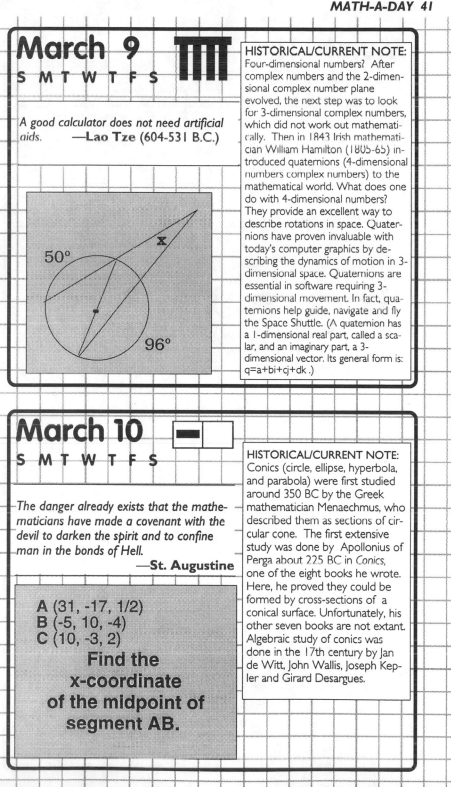

March 9
S M T W T F S

A good calculator does not need artificial aids. —**Lao Tze** (604-531 B.C.)

50°

x

96°

HISTORICAL/CURRENT NOTE:
Four-dimensional numbers? After complex numbers and the 2-dimensional complex number plane evolved, the next step was to look for 3-dimensional complex numbers, which did not work out mathematically. Then in 1843 Irish mathematician William Hamilton (1805-65) introduced quaternions (4-dimensional numbers complex numbers) to the mathematical world. What does one do with 4-dimensional numbers? They provide an excellent way to describe rotations in space. Quaternions have proven invaluable with today's computer graphics by describing the dynamics of motion in 3-dimensional space. Quaternions are essential in software requiring 3-dimensional movement. In fact, quaternions help guide, navigate and fly the Space Shuttle. (A quaternion has a 1-dimensional real part, called a scalar, and an imaginary part, a 3-dimensional vector. Its general form is: $q=a+bi+cj+dk$.)

March 10
S M T W T F S

The danger already exists that the mathematicians have made a covenant with the devil to darken the spirit and to confine man in the bonds of Hell.
—**St. Augustine**

A (31, -17, 1/2)
B (-5, 10, -4)
C (10, -3, 2)
**Find the
x-coordinate
of the midpoint of
segment AB.**

HISTORICAL/CURRENT NOTE:
Conics (circle, ellipse, hyperbola, and parabola) were first studied around 350 BC by the Greek mathematician Menaechmus, who described them as sections of circular cone. The first extensive study was done by Apollonius of Perga about 225 BC in *Conics*, one of the eight books he wrote. Here, he proved they could be formed by cross-sections of a conical surface. Unfortunately, his other seven books are not extant. Algebraic study of conics was done in the 17th century by Jan de Witt, John Wallis, Joseph Kepler and Girard Desargues.

March 11

S M T W T F S

There is no branch of mathematics, however abstract, which may not some day be applied to phenomena of the real world **—Nikolai Lobatchevsky**

$$\sum_{n=1}^{4}(x^2-x+1)$$

HISTORICAL/CURRENT NOTE:
In 1847 an algebraic system was devised by George Boole that revolutionized logical reasoning. His algebra, now called Boolean algebra, is able to reduce logical statements to a symbolic language in which reaching a conclusion or reasoning seems as mechanical as solving an algebraic equation. Logical statements are translated into the elements and binary operations of his system. Today this 19th century method has broad application in such fields as probability, topology and computer design.

March 12

S M T W T F S

It is rare to find learned men who are clean, do not stink and have a sense of humour. **—Gottfried Leibniz**

If the supplement of an angle is 6 more than 5 times the angle, then the angle's measure must be __?°__ .

HISTORICAL/CURRENT NOTE:
Among the women of antiquity Hypatia was the first famous mathematician (370 AD-415). Her father, Theon, a mathematician and the director of the University of Alexandria (called the Museum), encouraged her to pursue her scientific interest. Hypatia grew up in Alexandria, the most intellectual city of its time. She taught at the university, was famous for her engaging lectures, research, and beauty.

March 13 —|||

S M T W T F S

A great truth is a truth whose opposite is also a great truth. — **Thomas Mann**

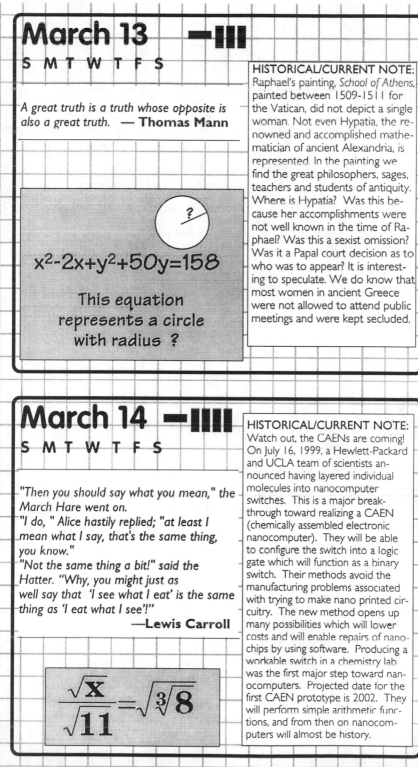

$$x^2-2x+y^2+50y=158$$

This equation represents a circle with radius **?**

HISTORICAL/CURRENT NOTE:
Raphael's painting, *School of Athens*, painted between 1509-1511 for the Vatican, did not depict a single woman. Not even Hypatia, the renowned and accomplished mathematician of ancient Alexandria, is represented. In the painting we find the great philosophers, sages, teachers and students of antiquity. Where is Hypatia? Was this because her accomplishments were not well known in the time of Raphael? Was this a sexist omission? Was it a Papal court decision as to who was to appear? It is interesting to speculate. We do know that most women in ancient Greece were not allowed to attend public meetings and were kept secluded.

March 14 —||||

S M T W T F S

"Then you should say what you mean," the March Hare went on.
"I do, " Alice hastily replied; "at least I mean what I say, that's the same thing, you know."
"Not the same thing a bit!" said the Hatter. "Why, you might just as well say that 'I see what I eat' is the same thing as 'I eat what I see'!"
—**Lewis Carroll**

$$\frac{\sqrt{x}}{\sqrt{11}}=\sqrt{\sqrt[3]{8}}$$

HISTORICAL/CURRENT NOTE:
Watch out, the CAENs are coming! On July 16, 1999, a Hewlett-Packard and UCLA team of scientists announced having layered individual molecules into nanocomputer switches. This is a major breakthrough toward realizing a CAEN (chemically assembled electronic nanocomputer). They will be able to configure the switch into a logic gate which will function as a binary switch. Their methods avoid the manufacturing problems associated with trying to make nano printed circuitry. The new method opens up many possibilities which will lower costs and will enable repairs of nanochips by using software. Producing a workable switch in a chemistry lab was the first major step toward nanocomputers. Projected date for the first CAEN prototype is 2002. They will perform simple arithmetic functions, and from then on nanocomputers will almost be history.

Giuseppe Peano intro-
duced this symbol into
set theory in 1889.
$x \in B$ means x belongs to
set B.

March 15 ▬❙❙❙❙❙

S M T W T F S

Euclid alone has looked on Beauty bare.
                    —Edna St. Vincent Millay

**HISTORICAL/CURRENT
NOTE:** Who took the bugs
out of Euclidean geometry? German mathematician David Hilbert. Today, it's is essentially his
modified Euclidean geometry,
based on his 1899 book *Foundations of Geometry,* that is taught in
schools.

$$\frac{\left(1.35 \times 10^{-3}\right)\left(8 \times 10^{15}\right)}{\left(2 \times 10^{-8}\right)\left(5.4 \times 10^{-6}\right)}$$

$$= 10^{?}$$

March 16 — **T**

S M T W T F S

I continued to do arithmetic with my father, passing proudly through fractions to decimals. I eventually arrived at the point where so many cows ate so much grass, and tanks filled with water in so many hours I found it quite enthralling.
— **Agatha Christie**

HISTORICAL/CURRENT NOTE:
How did Archimedes (circa 287-212 BC) find and show the surface area of a sphere was $4\pi r^2$?
He proved it was impossible for the sphere's surface area to be greater than or less than 4 times the area of a great circle of the sphere. That meant it had to equal 4 times the area of a great circle of the sphere.

If the 1st and 15th terms of an arithmetic sequence are 7 and 35 respectively, then what is the 8th term?

March 17 — **π**

S M T W T F S

God not only plays dice. He also sometimes throws the dice where they cannot be seen. — **Stephen Hawking**

If the radius of this sphere is $(3\sqrt{3\pi})/2\pi$, what is its surface area?

HISTORICAL/CURRENT NOTE:
Gerolamo Saccheri's (1591-1661) biggest mistake was not believing in what his mathematical work revealed. In 1639, he was the first to try to prove Euclid's Parallel postulate using an indirect proof. Prior to his death he published, *Euclides ab omni naeva vindicatus* (Euclid Freed of Every Flaw). One and a half a centuries later his book came to the attention of Eugenio Beltami, after non-Euclidean geometries had come to light in the 1800s. Had Saccheri not rejected his findings he probably would have sped up the discovery of a non-Euclidean geometry by about 100 years.

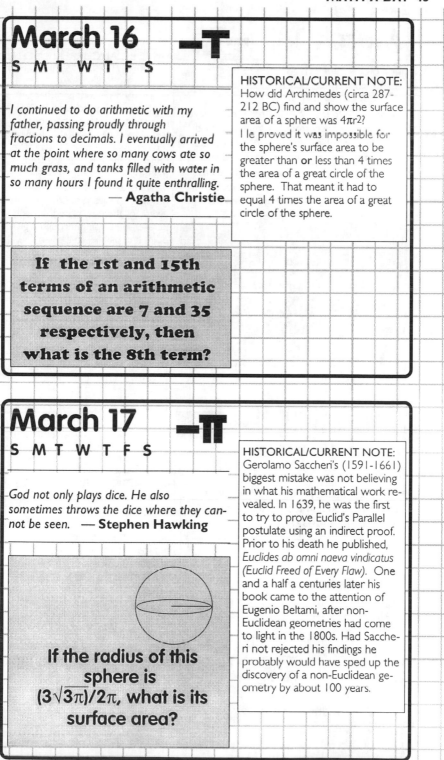

March 18 −ℼ

S M T W T F S

Don't talk to me of your Archimedes' lever. He was an absentminded person with a mathematical imagination. Mathematics commands all my respect, but I have no use for engines. Give me the right word and the right accent and I will move the world. —**Joseph Conrad**

$$\frac{-e^{\pi i}}{\log_2(\log_8 64)}$$

HISTORICAL/CURRENT NOTE: Felix Klein (1849-1925) christened Boyai/Lobachesky's non-Euclidean geometry *hyperbolic* and Riemann's *elliptic* geometry (today it is also referred to as spherical geometry). Klein's names stuck, and that is what they are called today. The origins of these words are from the Greek words *hyperbole*—meaning excessive and *elliptic*—meaning to fall short. In *hyperbolic geometry*, there is more than one parallel line passing through a given point and parallel to a given line, and *elliptic geometry* has no such parallel lines. If Klein had his way he might even have named Euclidean geometry *parabolic* geometry (parabolic meaning *to compare* with regards the distance between the parallel lines).

March 19 −ℼ

S M T W T F S

Logic, like whiskey, loses its beneficial effect when taken in too large quantities. —**Lord Dunsany**

The line-up problem

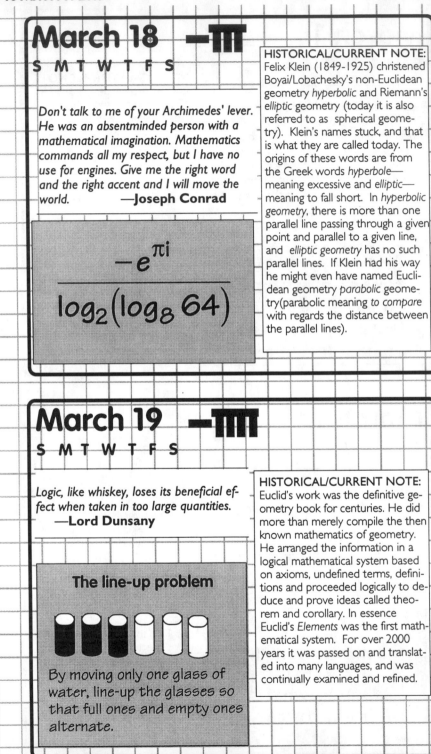

By moving only one glass of water, line-up the glasses so that full ones and empty ones alternate.

HISTORICAL/CURRENT NOTE: Euclid's work was the definitive geometry book for centuries. He did more than merely compile the then known mathematics of geometry. He arranged the information in a logical mathematical system based on axioms, undefined terms, definitions and proceeded logically to deduce and prove ideas called theorem and corollary. In essence Euclid's *Elements* was the first mathematical system. For over 2000 years it was passed on and translated into many languages, and was continually examined and refined.

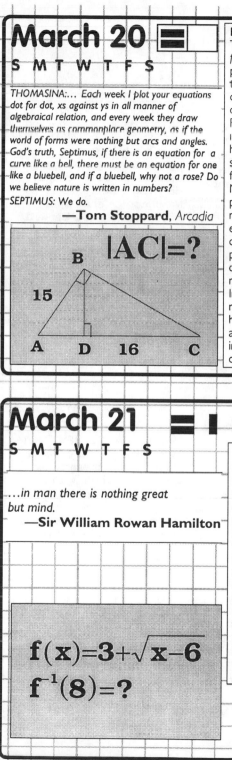

March 20

S M T W T F S

THOMASINA:... Each week I plot your equations dot for dot, xs against ys in all manner of algebraical relation, and every week they draw themselves as commonplace geometry, as if the world of forms were nothing but arcs and angles. God's truth, Septimus, if there is an equation for a curve like a bell, there must be an equation for one like a bluebell, and if a bluebell, why not a rose? Do we believe nature is written in numbers?

SEPTIMUS: We do.

— **Tom Stoppard**, Arcadia

$|AC|=?$

B

15

A D 16 C

HISTORICAL/CURRENT NOTE:
The first reference to the famous *four-color map problem* (proving any planar map requires only four colors to distinguish its boundaries) is one of those problems which grew out of a mathematical pastime. In 1852, Francis Guthrie, a graduate student in London shared the problem with his brother. Neither could solve it, so Guthrie passed it on to his professor, Augustus de Morgan. de Morgan couldn't prove it, and the problem began to spread in mathematical circles. Proofs were attempted, but all had errors. Some proofs delved into surfaces other than planes. This supposedly "simple" coloring problem evolved into a mammoth mathematical challenge. In 1976, Wolfgang Hallen and Kenneth Appel used a computer to exhaust all the necessary possibilities and present a proof which was impossible to verify without the aid of a computer.

March 21

S M T W T F S

...in man there is nothing great but mind.

— **Sir William Rowan Hamilton**

$$f(x)=3+\sqrt{x-6}$$

$$f^{-1}(8)=?$$

HISTORICAL/CURRENT NOTE:
In ancient times when people played the market, it was not the stock market. Thales (600 BC), one of the seven sages of the ancient world, was a shrewd business man who made a killing in the olive oil market. Yet, today his fame is not tied to his business acumen, but he is recognized as one of the first people to use deductive reasoning to prove mathematical ideas. Up until his time, people would accept mathematical ideas, such as the Pythagorean theorem, as true, if it worked out a number of times.

March 22

S M T W T F S

*The errors of definitions multiply them-
selves according as the reckoning proceeds; and lead men into absurdities,
which at last they see but cannot avoid,
without reckoning anew from the beginning.* **—Thomas Hobbes**

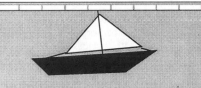

A boat heads out of the harbor
at 22° for 23 miles. It then
changes its bearing to 202° for
12 miles. How far is the boat
from the harbor at this point?

HISTORICAL/CURRENT NOTE:
What computer ushered in the
modern computer age? The Mark
I. Built in 1944 by Howard Hatha-
way Aiken, along with IBM engineers at Harvard University and
IBM headquarters, it was an enormous machine, especially by today's standards — 50' long and 8'
high with over a million parts. It
was the first digital computer that
could perform a variety of mathematical tasks.

March 23

S M T W T F S

*Statistics are the triumph of the quantita-
tive method, and the quantitative method
is the victory of sterility and death.*
—Hillaire Belloc

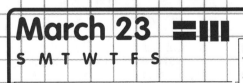

The hot water faucet fills
the tub in 80 minutes, while
the cold water faucet does
it in 48 minutes. If both are
turned on, how long would it
take to fill the tub?

HISTORICAL/CURRENT NOTE:
Why is Maria Agnesi always
associated with the phrase *the
witch of Agnesi* ? *The witch of
Agnesi* refers to the curve $xy^2=a^2$
$(a-x)$, which was one of numerous
topics she discussed in her book
*Instituzioni analitiche (Analytical
Institutions)*. This curve had also
been studied by earlier
mathematicians, among them
Guida Grandi and Fermat. They
referred to this curve as *versiera* (a
term from the Latin meaning *to
turn*). In 1801, when Englishman
John Colson translated Agnesi's
book, he mistakenly translated
versiera to be *witch*. Unfortunately,
this is the word immediately linked
with Agnesi whenever her name is
mentioned today.

March 24 ▰▮▮▮▮

S M T W T F S

Detection is, or ought to be, an exact science and should be treated in the same cold and unemotional manner. You have attempted to tinge it with romanticism, which produces much the same effect as if you worked a love story or an elopement into the fifth proposition of Euclid.

—Sir Arthur Conan Doyle

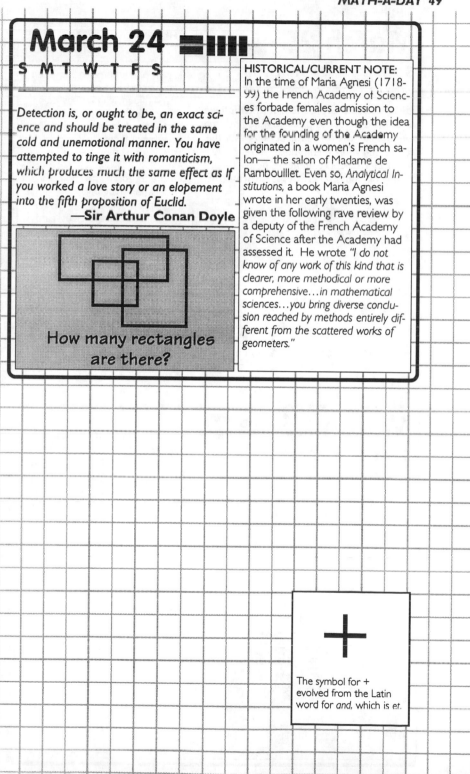

How many rectangles are there?

HISTORICAL/CURRENT NOTE:
In the time of Maria Agnesi (1718-99) the French Academy of Sciences forbade females admission to the Academy even though the idea for the founding of the Academy originated in a women's French salon— the salon of Madame de Rambouillet. Even so, *Analytical Institutions,* a book Maria Agnesi wrote in her early twenties, was given the following rave review by a deputy of the French Academy of Science after the Academy had assessed it. He wrote *"I do not know of any work of this kind that is clearer, more methodical or more comprehensive...in mathematical sciences...you bring diverse conclusion reached by methods entirely different from the scattered works of geometers."*

The symbol for + evolved from the Latin word for *and,* which is *et.*

March 25

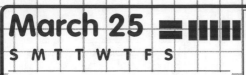

S M T W T F S

It is a pleasant surprise to him (the mathematician) and an added problem if he finds that the arts can use his calculations, or that the senses can verify them, much as if a composer found that sailors could heave better when singing his songs. —**George Santayana**

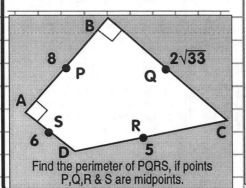

Find the perimeter of PQRS, if points P,Q,R & S are midpoints.

HISTORICAL/CURRENT NOTE:
By the age of nine, Maria Agnesi had mastered Greek, Latin, Hebrew plus a number of modern languages. Her parents were very influential in orchestrating their daughter's education, encouraging her interest and discussions of ideas. As early as the beginning of the Renaissance, Italy allowed its women more freedom and choices in education than any of the other European countries. Women could attend universities, earn degrees, become professors — in essence they were free to pursue their intellectual interests regardless of the field. It was in this climate that a mathematician such as Agnesi could flourish. It was here she was elected to the Academy of Bologna, receiving many accolades honoring her work in mathematics.

March 26

S M T W T F S

See skulking Truth to her old cavern fled,
Mountains of Casuistry heap'd o'er her head!
Philosophy, that lean'd on Heav'n before,
Shrinks to her second cause, and is no more.
Physic of Metaphysic begs defence,
And Metaphysic calls for aid on Sense!
See Mystery to Mathematics fly!
                              —**Alexander Pope**

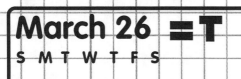

$$\sqrt{x\sqrt{x\sqrt{x\sqrt{x}}}}\ldots = 15$$

HISTORICAL/CURRENT NOTE:
In 1762, the University of Turin asked Maria Agnesi to review the work of the young mathematician Joseph Lagrange on the calculus of variation. There are not many instances in the history of mathematics where female mathematicians have been officially asked to critique the work of male mathematicians. At this point in her life Agnesi was devoting her time, energies, and wealth to the care of the poor, the ill, and others in dire need. She declined the offer.

March 27 = π

S M T W T F S

Where there is matter, there is geometry. —**Johannes Kepler**

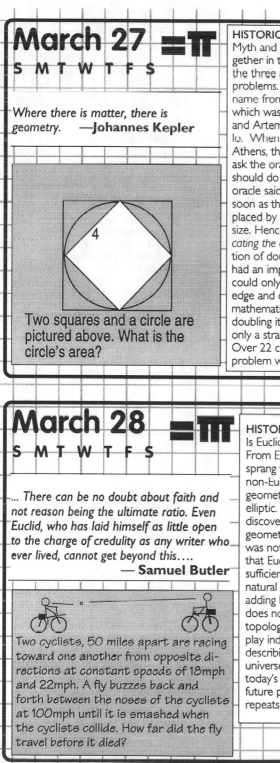

Two squares and a circle are pictured above. What is the circle's area?

HISTORICAL/CURRENT NOTE:
Myth and mathematics come together in the Delian problem, one of the three ancient famous impossible problems. The problem gets its name from the island of Delos, which was the birthplace of Apollo and Artemis, and the oracle of Apollo. When a plague was devastating Athens, the Athenians decided to ask the oracle of Apollo what they should do to stop the disease. The oracle said the plague would stop as soon as the altar of Apollo was replaced by one exactly double its size. Hence the origin of the *duplicating the cube* problem. The solution of doubling the cube's volume had an important restriction— it could only be done using a straight-edge and compass. The ancient mathematicians devised ways of doubling its volume, but none using only a straight-edge and compass. Over 22 centuries later the Delian problem was proven impossible.

March 28 = π

S M T W T F S

... There can be no doubt about faith and not reason being the ultimate ratio. Even Euclid, who has laid himself as little open to the charge of credulity as any writer who ever lived, cannot get beyond this....
— **Samuel Butler**

Two cyclists, 50 miles apart are racing toward one another from opposite directions at constant speeds of 18mph and 22mph. A fly buzzes back and forth between the noses of the cyclists at 100mph until it is smashed when the cyclists collide. How far did the fly travel before it died?

HISTORICAL/CURRENT NOTE:
Is Euclidean geometry enough? From Euclid's Parallel Postulate sprang two amazing non-Euclidean geometries—hyperbolic and elliptic. At the time of the discoveries of non-Euclidean geometries, their immediate use was not obvious. Today, we see that Euclidean geometry is not sufficient to describe and explain natural phenomena. In fact, adding hyperbolic and elliptic does not suffice. Others such as topology and fractal geometry play indispensable roles in describing and analyzing the universe. Will the geometries of today's mathematics suffice for future problems? If history repeats itself, probably not.

March 29 ▬▬▬▬▬TTTT

S M T W T F S

Finally I am becoming stupider no more.
—Paul Erdös
the epitaph he wrote for himself

Without lifting your pencil connect the 9 dots with just 4 straight lines.

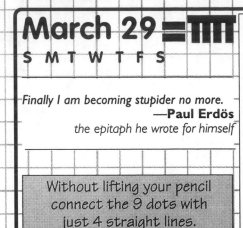

HISTORICAL/CURRENT NOTE:
Playing the dumb broad was no joke in France. During the time of Louis XIV, the first state school for girls, the Institut de St.-Cyr, was founded. Unfortunately, it was only accessible to nobility and taught subjects superficially. Breaking with tradition and societal pressures, Emile de Breteuil (Marquise du Chatelet) studied mathematics with a passion. She was introduced by Voltaire to physicists who discussed Newton's work. She had to resort to tutors since advanced schools were not available to her. She would work for hours on mathematics and Newtonian physics. In fact, she secretly wrote her own book on physics that incorporated the works of Newton and Leibniz. When her tutor, Samuel König, got wind of this, he was concerned that she would outshine him, and he spread the word that the work was really his. This lie was accepted, since such work was considered outside a woman's realm. Consequently, she had to publish the work anonymously. Today, Chatelet is given credit for introducing Newton to the French by having made the first French translation of Newton's *Principia*.

March 30 ▬▬

S M T W T F S

It is a mathematical fact that the casting of this pebble from my hand alters the centre of gravity of the universe
—Thomas Carlyle

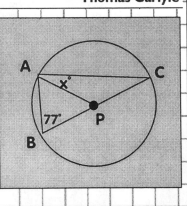

HISTORICAL/CURRENT NOTE:
In the early 1600s, Galileo began experimenting with telescopes. He eventually constructed one with magnification power 33 times, which, like the Hubble today, brought into view things never before seen in the night sky. Galileo saw the 3 moons of Jupiter, and noticed that they circled Jupiter. These observations challenged the accepted geocentric view of the solar system, in which everything revolved around the Earth. Since Jupiter had moons, did this imply the Earth must also be a planet, since it had a moon? Since Jupiter's moons revolved around Jupiter and not the Earth, was the Earth not the center of the universe? Thus Galileo, the telescope and what it revealed placed the prevailing theory of the universe on unsteady ground. It wasn't the first time the geocentric theory had been questioned. In the 3rd century BC, Greek astronomer Aristarchus of Samos contended that the planets revolved around the Sun, but this idea was ignored probably because other Greeks, such as Aristotle and Ptolemy, firmly believed in the geocentric idea. Copernicus' theory, mathematics, new technology and Galileo's careful observations proved geocentrism false. In 1999, the Roman Catholic Church acknowledged their error in forcing Galileo to renounce what he knew was true, thereby vindicating Galileo.

Leibniz is responsible for the symbolic notation used in calculus, but for multiplication he preferred this symbol rather than ✕. He felt ✕ could easily be confused with algebra's symbol for "unknown X".

March 31

S M T W T F S

The mathematician may be compared to a designer of garments, who is utterly oblivious of the creatures whom his garments may fit. —**Tobias Dantzig**

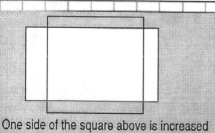

One side of the square above is increased by 3 units while its adjacent side is decreased by 2 units. The resulting rectangle has perimeter 26 units. What is the length of the rectangle?

HISTORICAL/CURRENT NOTE: In mathematics the word *cut* has a totally different significance than its English language counterpart. From ancient times, irrational numbers were loosely described as those that could not be expressed as fractions. Later, they were considered never ending non-repeating decimals. But with the introduction of the simple word *cut*, Julius Dedekind (1831-1916) was able to define an irrational number by looking at two sets of rational numbers with respect to the irrational numbers. One set does not have a largest member and the other does not have a smallest member. For example, to define √3, set A is described as having all rational numbers with squares less than 3, and set B is described as having all rational number with squares greater than 3. His ingenious method not only defined irrationals, but filled in the holes on the real number line occupied by the irrationals.

April 1   A

April dates are in the Greek aphabet numerals, also known as the Ionic number system

S M T W T F S

Mathematics is the tool specially suited for dealing with abstract concepts of any kind and there is no limit to its power in this field.   **—Paul Dirac**

HISTORICAL/CURRENT NOTE: Nanotechnology—the world of the ultra small— has entered the mainstream. In 1999, the White House called nanotechnology the technology of the 21st century. In addition, a Congressional proposal seeks to double the Federal Government's annual spending on nanotech research. Spending in 1999 on nanotech research totaled about 233 million.

Which of the shapes below are not practical to use for manholes, and why?

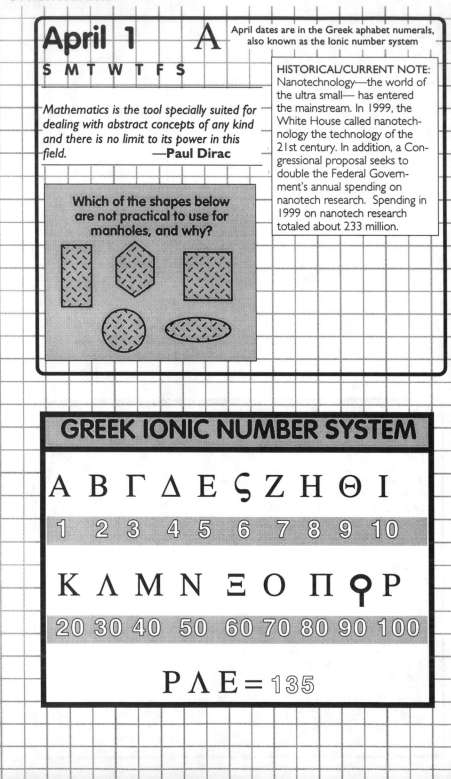

GREEK IONIC NUMBER SYSTEM

A	B	Γ	Δ	E	Ϛ	Z	H	Θ	I
1	2	3	4	5	6	7	8	9	10

K	Λ	M	N	Ξ	O	Π	Ϙ	P
20	30	40	50	60	70	80	90	100

$$P \Lambda E = 135$$

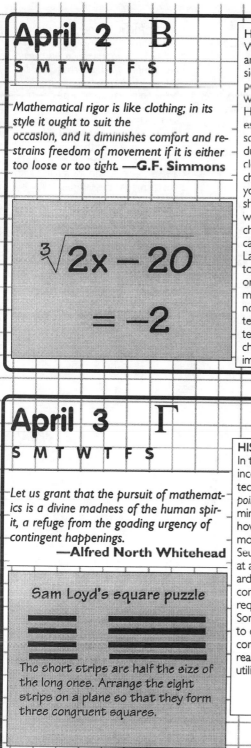

April 2 B

S M T W T F S

Mathematical rigor is like clothing; in its style it ought to suit the occasion, and it diminishes comfort and restrains freedom of movement if it is either too loose or too tight. **—G.F. Simmons**

$$\sqrt[3]{2x - 20}$$

$$= -2$$

HISTORICAL/CURRENT NOTE: What kind of commercial products are nanotech entrepeuners envisioning? Try programmable nail polish that paints the designs you want when applied to your nails. How about housepaint that changes color when you want to try something different. The textile industry wants temperature sensitive clothing that adapts to the climatic changes. How about wrapping your sandwich with voice activated shrink wrap? The airplane industry would like smart metals that change the make-up of their fabrication according to air conditions. Landscapers want nanosnail baiters to destroy those pesky creatures, or soil that senses and adapts moisture for your plants— imagine not having to worry about a watering schedule. If nanoworld materializes, we're in for major changes limited only by our imaginations.

April 3 Γ

S M T W T F S

Let us grant that the pursuit of mathematics is a divine madness of the human spirit, a refuge from the goading urgency of contingent happenings.
—Alfred North Whitehead

Sam Loyd's square puzzle

The short strips are half the size of the long ones. Arrange the eight strips on a plane so that they form three congruent squares.

HISTORICAL/CURRENT NOTE: In the mid 1800s Georges Seurat incorporated impressionistic techniques in a new style called *pointillism*. Pointillism employed minute dots of paint, analogous to how each pixel of a computer monitor is a painted square. Seurat's dots blend when viewed at a distance. His art was an arduous process similar to solving a complex mathematical problem requiring following sequential steps. Some of his works took him years to complete. Today, the pixel and computer technology create realistic images through software utilizing mathematical formulas.

April 4 △

S M T W T F S

Yes, yes, I know that Sidney…everybody knows that!…But look: *Four wrongs squared, minus two wrongs to the fourth power, divided by this formula, do make a right.* **—Gary Larson**

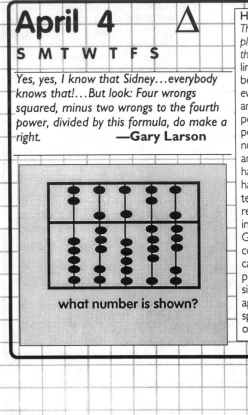

what number is shown?

HISTORICAL/CURRENT NOTE:
This old man he played 4, he played… What mystique lies behind the number 4? The Pythagoreans linked 4 to harmony and justice because it was composed of the even-even number, namely 2+2=4 and 2•2=4. Being the first composite number and and the first perfect square of the first prime number gave it special status in the ancient world. The mystique of 4 has grown as the famous 4s have— the 4 elements (earth, water, air, and fire), the 4 compass directions, the 4 rivers of paradise in the Old Testament, the 4 Gospels of the New Testament, 4 corners of the world. Every integer can be expressed by at most 4 perfect squares, and the dimensions of space-time are 4. 4 appears in measuring with one span being 4 fingers, and 4 breaths of a palm called a foot.

The division symbol first appeared in print in a book by Johann Heinrich Rahn in 1659.

April 5   E

S M T W T F S

As for everything else, so for a mathematical theory: beauty can be perceived but not explained.
—Arthur Cayley

$$A(-3, -9\sqrt{11})$$
$$B(7, -7\sqrt{11})$$
$$|AB| = ?$$

HISTORICAL/CURRENT NOTE:
What's a *one-to-one correspondence*? It matches one object from one set of things with only one from another set of things, and *vice versa*. One of the most ingenious one-to-one correspondences ever devised was how each rational number was matched with only one counting number, and *vice versa*. At a glance, you'd think that would not be possible because there are infinitely many rational numbers between any two counting numbers. For example, between 1 and 2 we find 1 1/2, 1 1/3, 1 1/4, 1 1/5, and so on. Yet, the mathematical genius Georg Cantor (1845-1919) discovered a way to match them up and launch his ground breaking work on set theory and transfinite numbers.

April 6   Ç

S M T W T F S

In many cases, mathematics is an escape from reality. The mathematician finds his own monastic niche and happiness in pursuits that are disconnected from external affairs.... In their unhappiness over the events of this world, some immerse themselves in a kind of self-sufficiency in mathematics. **—Stanislaw Ulam**

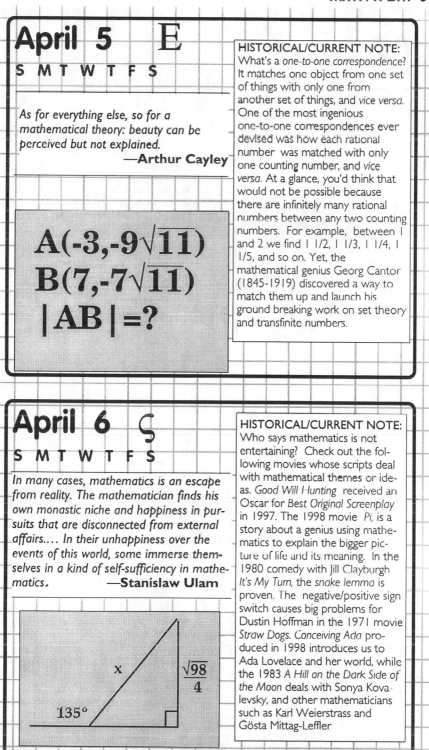

HISTORICAL/CURRENT NOTE:
Who says mathematics is not entertaining? Check out the following movies whose scripts deal with mathematical themes or ideas. *Good Will Hunting* received an Oscar for *Best Original Screenplay* in 1997. The 1998 movie *Pi*, is a story about a genius using mathematics to explain the bigger picture of life and its meaning. In the 1980 comedy with Jill Clayburgh *It's My Turn*, the *snake lemma* is proven. The negative/positive sign switch causes big problems for Dustin Hoffman in the 1971 movie *Straw Dogs. Conceiving Ada* produced in 1998 introduces us to Ada Lovelace and her world, while the 1983 *A Hill on the Dark Side of the Moon* deals with Sonya Kovalevsky, and other mathematicians such as Karl Weierstrass and Gösta Mittag-Leffler

April 7 Z
S M T T W T F S

Obvious is the most dangerous word in mathematics. —**Eric Temple Bell**

What is the positive root for $5x^2 + 20x = 105$?

April 8 H
S M T W T F S

All the effects of nature are but the mathematical results of a small number of immutable laws.
—**Pierre-Simon de Laplace**

For $a_{n+2} = 2a_{n+1} + a_n$ $n \in \{\text{natural numbers}\}$, which term has value 219, if $a_1 = 3$ & $a_2 = 17$?

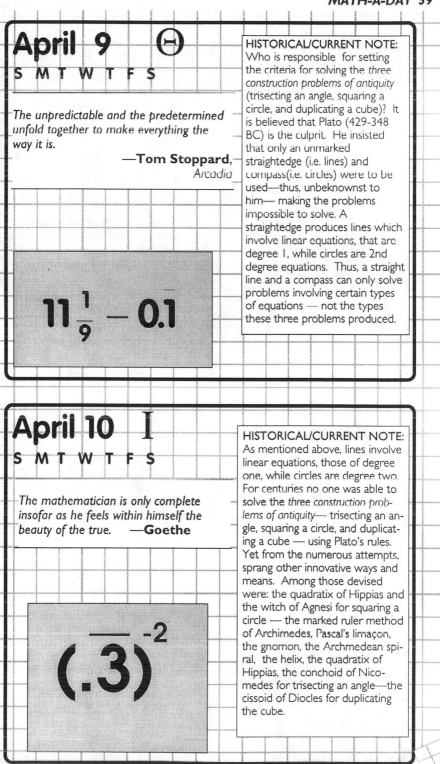

April 9 ⊖

S M T W T F S

The unpredictable and the predetermined unfold together to make everything the way it is.

—Tom Stoppard,
Arcadia

$$11\frac{1}{9} - 0.1$$

HISTORICAL/CURRENT NOTE:
Who is responsible for setting the criteria for solving the *three construction problems of antiquity* (trisecting an angle, squaring a circle, and duplicating a cube)? It is believed that Plato (429-348 BC) is the culprit. He insisted that only an unmarked straightedge (i.e. lines) and compass (i.e. circles) were to be used—thus, unbeknownst to him— making the problems impossible to solve. A straightedge produces lines which involve linear equations, that are degree 1, while circles are 2nd degree equations. Thus, a straight line and a compass can only solve problems involving certain types of equations — not the types these three problems produced.

April 10 I

S M T W T F S

The mathematician is only complete insofar as he feels within himself the beauty of the true. **—Goethe**

$$(.\overline{3})^{-2}$$

HISTORICAL/CURRENT NOTE:
As mentioned above, lines involve linear equations, those of degree one, while circles are degree two. For centuries no one was able to solve the *three construction problems of antiquity*— trisecting an angle, squaring a circle, and duplicating a cube — using Plato's rules. Yet from the numerous attempts, sprang other innovative ways and means. Among those devised were: the quadratix of Hippias and the witch of Agnesi for squaring a circle — the marked ruler method of Archimedes, Pascal's limaçon, the gnomon, the Archimedean spiral, the helix, the quadratix of Hippias, the conchoid of Nicomedes for trisecting an angle—the cissoid of Diocles for duplicating the cube.

April 11   IA

S M T W T F S

What's…more powerful than God,
more evil than the devil;
the poor have it,
the rich lack it,
and if you eat it you die?
        —Margaret Atwood,
                The Blind Assassin

$$18+6+2+\frac{2}{3}+$$
$$\frac{2}{9}+\frac{2}{27}+\frac{2}{54}+\ldots=\,?$$

HISTORICAL/CURRENT NOTE:
Many people read their daily horoscope. Even though Joahnnes Kepler considered astrology foolish,
Kepler did compile horoscopes for
some members of the nobility
who requested them. Why? Perhaps as a favor, or perhaps to
make ends meet. Regardless, one
of those horoscopes has survived.
The 16th century old horoscope,
which Kepler wrote for Hans Han-
nibal Hutter von Hutterhofen, was
discovered in 1998 by Anthony
Misch in the Mary Lea Shane Ar-
chives of the University of California at Santa Cruz in the McHenry
Library, after being lost for nearly a
century. A copy of the manuscript
is on display. This fascinating manuscript somehow made it from the
Pulkuva Observatory outside of St.
Peterburg , Russia, then to the Lick
Observatory, and finally to the
McHenry Library.

April 12   IB

S M T W T F S

As far as the laws of mathematics refer to
reality, they are not certain; and as far as
they are certain, they do not refer to
reality.   **—Albert Einstein**

AN OLD CLASSIC PROBLEM
A farmer wants to get his goat, wolf and
cabbage to the other side of the river. His boat
is only big enough to carry him and either his
goat, his wolf or his cabbage. If he leaves the
goat alone with the cabbage, the goat will eat
the cabbage. If he leaves the wolf with the goat,
the wolf will eat the goat. Only when the farmer
is present are the goat and the cabbage safe
from their predators. How does he manage to
get everything to the other side?

HISTORICAL/CURRENT NOTE:
TWINKLE (**t**he **W**eizmann
Institute **k**ey **l**ocating **e**ngine) is a
device, which when built with its
twinkling light emitting diodes, will
be able to factor numbers as
large as 150 digits long to its
prime factors. TWINKLE will
work as fast as thousands of
computers working
simultaneously on factoring a
number, and thereby put in
jeopardy the RSA encryption
system used to protect
information transmitted along the
Internet. The ironic twist is that
TWINKLE's designer is Adi
Shamir who was one of the
creators of the RSA encryption
system.

April 13 IΓ

S M T W T F S

Geometry is the only science that it hath pleased God hitherto to bestow on mankind.

— **Thomas Hobbes**

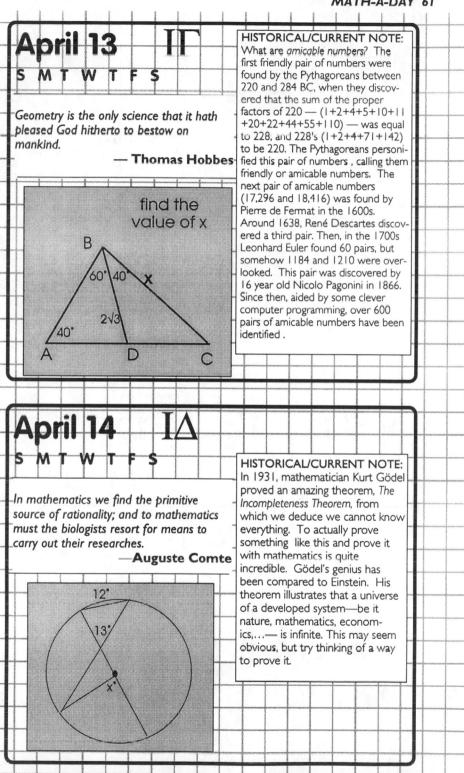

find the value of x

HISTORICAL/CURRENT NOTE:
What are *amicable numbers?* The first friendly pair of numbers were found by the Pythagoreans between 220 and 284 BC, when they discovered that the sum of the proper factors of 220 — (1+2+4+5+10+11 +20+22+44+55+110) — was equal to 228, and 228's (1+2+4+71+142) to be 220. The Pythagoreans personified this pair of numbers, calling them friendly or amicable numbers. The next pair of amicable numbers (17,296 and 18,416) was found by Pierre de Fermat in the 1600s. Around 1638, René Descartes discovered a third pair. Then, in the 1700s Leonhard Euler found 60 pairs, but somehow 1184 and 1210 were overlooked. This pair was discovered by 16 year old Nicolo Pagonini in 1866. Since then, aided by some clever computer programming, over 600 pairs of amicable numbers have been identified.

April 14 IΔ

S M T W T F S

In mathematics we find the primitive source of rationality; and to mathematics must the biologists resort for means to carry out their researches.

—**Auguste Comte**

HISTORICAL/CURRENT NOTE:
In 1931, mathematician Kurt Gödel proved an amazing theorem, *The Incompleteness Theorem,* from which we deduce we cannot know everything. To actually prove something like this and prove it with mathematics is quite incredible. Gödel's genius has been compared to Einstein. His theorem illustrates that a universe of a developed system—be it nature, mathematics, economics,...— is infinite. This may seem obvious, but try thinking of a way to prove it.

X

William Oughtred introduced the use of this symbol for multiplication in 1631. It had previously been used in dividing fractions by *cross multiplication*.

April 15   IE

S M T W T F S

If you can measure that of which you speak, and can express it by a number, you know something of your subject; but if you cannot measure it, your knowledge is meager and unsatisfactory.
—**William Thomson, Lord Kelvin**

$$f(x) = x^3 - 4x$$

$$\text{find } f'(2\sqrt{2})$$

HISTORICAL/CURRENT NOTE: Just as Bertrand Russell's paradox made the foundation of Gottlob Frege's work collapse, mathematician Kurt Gödel turned the tables on Russell with his *Incompleteness Theorem*. Gödel shook the mathematical world with his announcement in 1931 of his *Incompleteness Theorem* — in which he proved *a* consistent mathematical system formed from a finite number of axioms for deducing other ideas from them will have at least one true statement which the system cannot prove. This theorem put the kabash on Alfred Whitehead and Bertrand Russell's work *Principia Mathematica*. Their consistent and complete mathematical system was no longer complete.

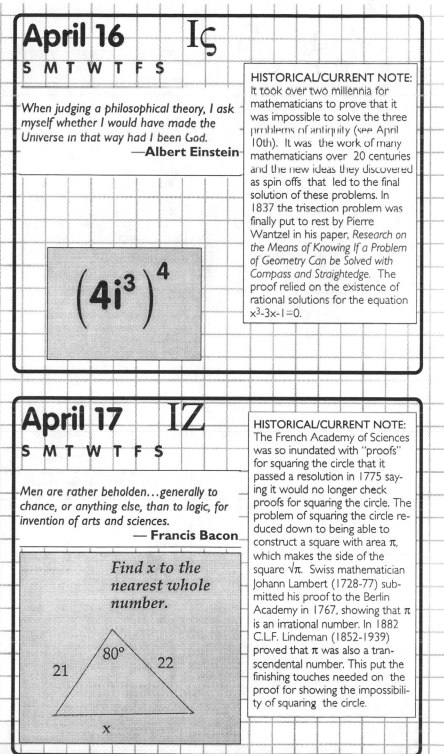

April 16   Iς

S M T W T F S

When judging a philosophical theory, I ask myself whether I would have made the Universe in that way had I been God.
—Albert Einstein

$$\left(4i^3\right)^4$$

HISTORICAL/CURRENT NOTE:
It took over two millennia for mathematicians to prove that it was impossible to solve the three problems of antiquity (see April 10th). It was the work of many mathematicians over 20 centuries and the new ideas they discovered as spin offs that led to the final solution of these problems. In 1837 the trisection problem was finally put to rest by Pierre Wantzel in his paper, *Research on the Means of Knowing If a Problem of Geometry Can be Solved with Compass and Straightedge.* The proof relied on the existence of rational solutions for the equation $x^3-3x-1=0$.

April 17   IZ

S M T W T F S

Men are rather beholden...generally to chance, or anything else, than to logic, for invention of arts and sciences.
— Francis Bacon

Find x to the nearest whole number.

80°
21
22
x

HISTORICAL/CURRENT NOTE:
The French Academy of Sciences was so inundated with "proofs" for squaring the circle that it passed a resolution in 1775 saying it would no longer check proofs for squaring the circle. The problem of squaring the circle reduced down to being able to construct a square with area π, which makes the side of the square $\sqrt{\pi}$. Swiss mathematician Johann Lambert (1728-77) submitted his proof to the Berlin Academy in 1767, showing that π is an irrational number. In 1882 C.L.F. Lindeman (1852-1939) proved that π was also a transcendental number. This put the finishing touches needed on the proof for showing the impossibility of squaring the circle.

April 18 IH

S M T W T F S

Statistical thinking will one day be as necessary for efficient citizenship as the ability to read and write. — **H.G. Wells**

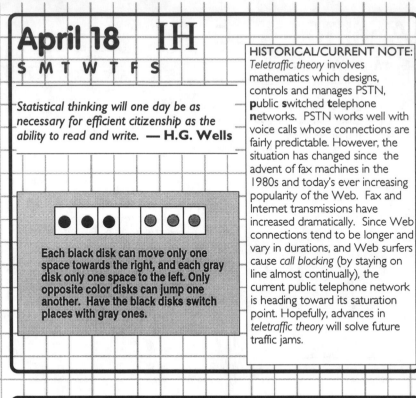

Each black disk can move only one space towards the right, and each gray disk only one space to the left. Only opposite color disks can jump one another. Have the black disks switch places with gray ones.

HISTORICAL/CURRENT NOTE:
Teletraffic theory involves mathematics which designs, controls and manages PSTN, **p**ublic **s**witched **t**elephone **n**etworks. PSTN works well with voice calls whose connections are fairly predictable. However, the situation has changed since the advent of fax machines in the 1980s and today's ever increasing popularity of the Web. Fax and Internet transmissions have increased dramatically. Since Web connections tend to be longer and vary in durations, and Web surfers cause *call blocking* (by staying on line almost continually), the current public telephone network is heading toward its saturation point. Hopefully, advances in *teletraffic theory* will solve future traffic jams.

April 19 IΘ

S M T W T F S

A good puzzle should demand the exercise of our best wit and ingenuity, and although a knowledge of mathematics...and...of logic are often of great service in the solution of these things, yet it sometimes happens that a kind of natural cunning and sagacity is of considerable value.—**Henry E. Dudeny**

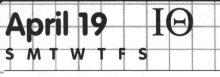

$$\left[(.4)\left(\frac{5}{8}\right)6!\right] \div 10$$

HISTORICAL/CURRENT NOTE:
Does your message get across? For Web transmission can you always connect? On the Web, individual *datagrams* or *data packets* are independently transmitted over the network. Each is self-contained with a header that identifies its destination. No inventory is needed. The routers simply use the headers to direct it through the net. This *packet-switching system* does away with the need for circuit-switching, since routers do not have to keep track of active connections. Each packet of data vies with others for a path. If there is no traffic, a packet can have a *bandwidth* to itself, and zip along at breakneck speed. If there's a lot of competition along the same path the bandwidth is shared. When packets exceed a link's capacity, the link is overloaded. Packets then get *buffered* (await transmission along the link). If the condition lasts too long, *congestion* sets in, and buffers in routes fill up and packets are discarded.

April 20 K

S M T T W T F S

Not chaos-like, together crushed
   and bruised,
But, as the world harmoniously confused;
Where order in variety we see,
And where, though all things differ,
   all agree.

—**Alexander Pope**

$$\log_x \text{million} = 6$$

HISTORICAL/CURRENT NOTE:
The ancient mathematician Pappus (circa 320 AD) claimed (without proof) in his writings that the *three construction problems of antiquity* were not possible by just using only a compass and straightedge. These proofs were not completely formulated until the 19th century.

April 21 KA

S M T W T F S

I have hardly ever known a mathematician who was capable of reasoning. —**Plato**

Find the area
of this
trapezoid in
square meters.

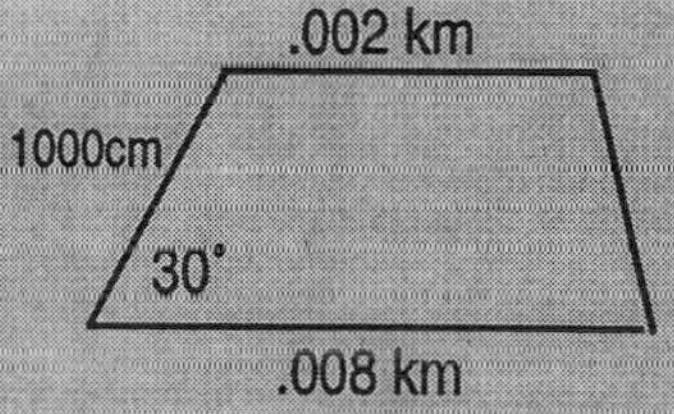

.002 km

1000cm

30°

.008 km

HISTORICAL/CURRENT NOTE:
Ever notice your modem is not allowed to work at its capacity on the Net? What causes this? It is how the Net deals with traffic *congestion.* An *end-to-end congestion control* mechanism kicks in and slows things down by automatically decreasing the transmission rate. Also built into *congestion control* is history— these are patterns each link has encountered in its past. The Internet traffic control correlates time and space and link interaction. Next time you can't get on line for awhile, *end-to-end congestion control* has booted your packet of data off line.

April 22 KB

S M T W T F S

I have resolved to quit abstract geometry...in order to study another kind of geometry, which has for its objects the explanation of the phenomena of nature.
—**René Descartes**

What number is this Babylonian number when written in Hindu-Arabic numerals?

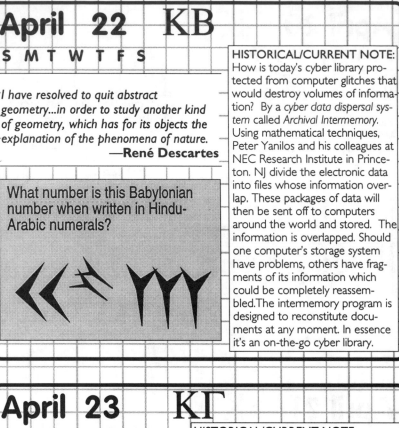

HISTORICAL/CURRENT NOTE:
How is today's cyber library protected from computer glitches that would destroy volumes of information? By a *cyber data dispersal system* called *Archival Intermemory*. Using mathematical techniques, Peter Yanilos and his colleagues at NEC Research Institute in Princeton. NJ divide the electronic data into files whose information overlap. These packages of data will then be sent off to computers around the world and stored. The information is overlapped. Should one computer's storage system have problems, others have fragments of its information which could be completely reassembled.The intermemory program is designed to reconstitute documents at any moment. In essence it's an on-the-go cyber library.

April 23 KΓ

S M T W T F S

There is nothing so troublesome to mathematical practice...than multiplications, divisions, square and cubical extractions of great numbers... I began therefore to consider...how I might remove those hindrances.
— **John Napier**

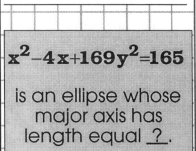

$$x^2 - 4x + 169y^2 = 165$$

is an ellipse whose major axis has length equal _?_ .

HISTORICAL/CURRENT NOTE:
Emmy Noether was born in 1882 in southern Germany. Her father, a mathematics professor, encouraged her passion for mathematics. Had she not been a woman, her talents would have allowed her to excel in the mathematical world. Yet, she lived in times when ignomace, racism, and sexism were predominant. Being a Jew, an intellectual liberal and a female were all against her, especially during post World War I and later during Nazi Germany. Unlike many mathematicians, who do their best work during their youth, Noether made outstanding contributions to mathematics during her entire life. Her ideas and thoughts were highly respected by such eminent mathematicians as David Hilbert, Felix Klein and Herman Weyl. Albert Einstein said "...*Fraulein Noether was the most significant creative mathematical genius thus far produced since the higher education of women began. In the realm of algebra...she discovered methods which have proved of enormous importance in the development of the present day younger generation of mathematicians.*" (New York Times, 5/4/1935) She was able to leave Nazi Germany and get a teaching position at Bryn Mawr after coming to the United States. Her work and success in the United States were impressive, but, unfortunately, she died one and a half years later, at the age of 53, from complications of an operation.

April 24 KΔ

S M T W T F S

Mathematics...possesses a beauty cold and austere...yet sublimely pure, and capable of a stern perfection such as only the greatest art can show. —**Bertrand Russell**

$$\frac{25\pi}{6}$$

has reference angle = ? degrees

HISTORICAL/CURRENT NOTE:
Did the ancient Greeks have a method for writing algebraic expresssions? Even though the Greeks are known predominantly for their geometry, Diophantus developed algebraic symbolism which he used to write his Diophantine equations. He wrote the coefficients of an unknown after the known. The coefficient would have a line drawn over it. So $\Delta\,\overline{\gamma}$ would mean 3x. All terms that were subtracted were written after the his subtraction symbol, Λ. Terms added were written side by side with nothing between them, and the same was done with terms being subtracted. Using the table above, the expression
$2x^3-x^2-5x+3$ would have been written as

$$K^Y\overline{\beta}\,M\,\overline{\gamma}\,\Lambda\,\varsigma\,\overline{\varepsilon}\,\Delta^Y\overline{\alpha}$$

Symbols that Diophantus used to write his equations

M meant $1\,x^0$

ς meant x or x^1

Δ^Y meant x^2

K^Y meant x^3

$\Delta^Y\Delta$ meant square-square, x^4

ΔK^Y meant square-cube x^5

K^Y meant cube-cube x^6

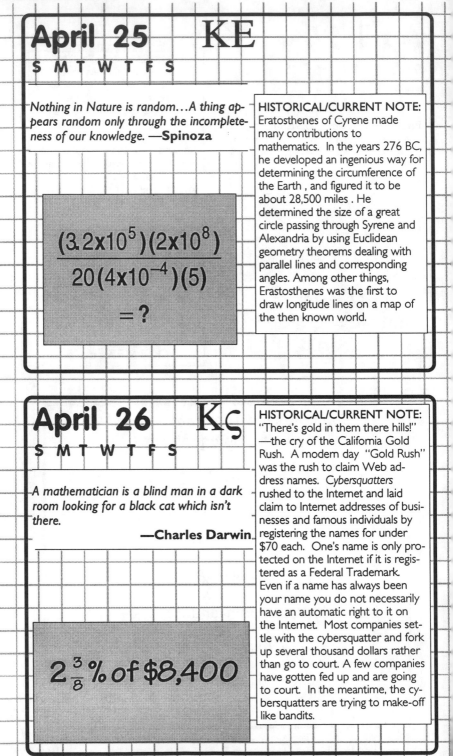

April 25   KE

S M T W T F S

Nothing in Nature is random...A thing appears random only through the incompleteness of our knowledge. —**Spinoza**

$$\frac{(3.2 \times 10^5)(2 \times 10^8)}{20(4 \times 10^{-4})(5)}$$

$$= ?$$

HISTORICAL/CURRENT NOTE:
Eratosthenes of Cyrene made many contributions to mathematics. In the years 276 BC, he developed an ingenious way for determining the circumference of the Earth , and figured it to be about 28,500 miles . He determined the size of a great circle passing through Syrene and Alexandria by using Euclidean geometry theorems dealing with parallel lines and corresponding angles. Among other things, Erastosthenes was the first to draw longitude lines on a map of the then known world.

April 26   Kς

S M T W T F S

A mathematician is a blind man in a dark room looking for a black cat which isn't there.

—**Charles Darwin**

$$2\frac{3}{8}\% \text{ of } \$8,400$$

HISTORICAL/CURRENT NOTE:
"There's gold in them there hills!" —the cry of the Califomia Gold Rush. A modem day "Gold Rush" was the rush to claim Web address names. *Cybersquatters* rushed to the Internet and laid claim to Internet addresses of businesses and famous individuals by registering the names for under $70 each. One's name is only protected on the Internet if it is registered as a Federal Trademark. Even if a name has always been your name you do not necessarily have an automatic right to it on the Internet. Most companies settle with the cybersquatter and fork up several thousand dollars rather than go to court. A few companies have gotten fed up and are going to court. In the meantime, the cybersquatters are trying to make-off like bandits.

April 27 KZ

S M T W T F S

I have often admired the mystical way of Pythagoras, and the secret magic of numbers. —**Sir Thomas Browne**

$$y = 5\sin\left(2\theta + 90^\circ\right)$$

graph this
sinus curve

HISTORICAL/CURRENT NOTE:
From ancient times, people have used codes to communicate things they wanted to keep secret. Ingenious methods and codes have been devised over the centuries, from the ancient Spartan scytale and leather strip to the infamous German *Engima* code of World War II. One simple, yet very resourceful code tool — the tapping code— was used by prisoners of war to enable them to communicate with one another. Dropping the letter K from the English alphabet leaves 25 letters which are arranged in an array of five letters

A	B	C	D	E
F	G	H	I	J
L	M	N	O	P
Q	R	S	T	U
V	W	X	Y	Z

and five rows. The prisoners could not use the Morse code because there was no way to tap a dash. Instead the array of letters was an easy solution. To tap an H, 2 taps with a pause gives the row followed by 3 taps gave the letter in that row, namely H.

April 28 KH

S M T W T F S

'Mathematizing' may well be a creative activity of man, like language or music, of primary orginality...
— **Herman Weyl**

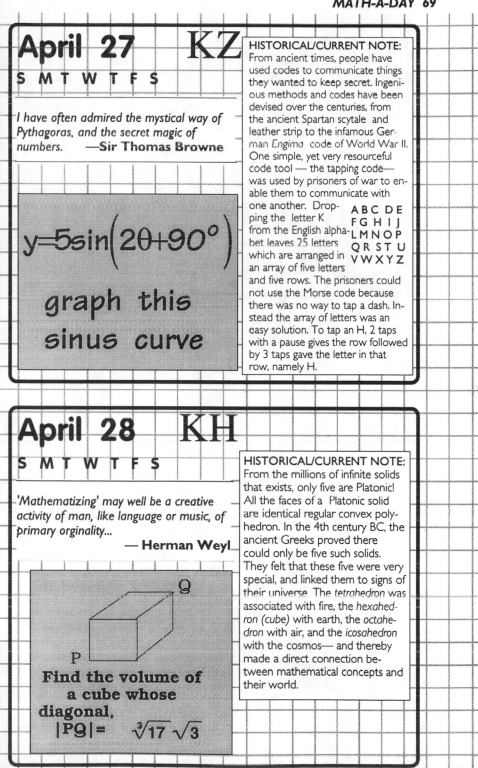

Find the volume of a cube whose diagonal,
$$|PQ| = \sqrt[3]{17}\sqrt{3}$$

HISTORICAL/CURRENT NOTE:
From the millions of infinite solids that exists, only five are Platonic! All the faces of a Platonic solid are identical regular convex polyhedron. In the 4th century BC, the ancient Greeks proved there could only be five such solids. They felt that these five were very special, and linked them to signs of their universe. The *tetrahedron* was associated with fire, the *hexahedron (cube)* with earth, the *octahedron* with air, and the *icosahedron* with the cosmos— and thereby made a direct connection between mathematical concepts and their world.

April 29 KΘ

S M T W T F S

A mathematician, like a painter or poet, is a maker of patterns...with ideas.
—**Godfrey H. Hardy**

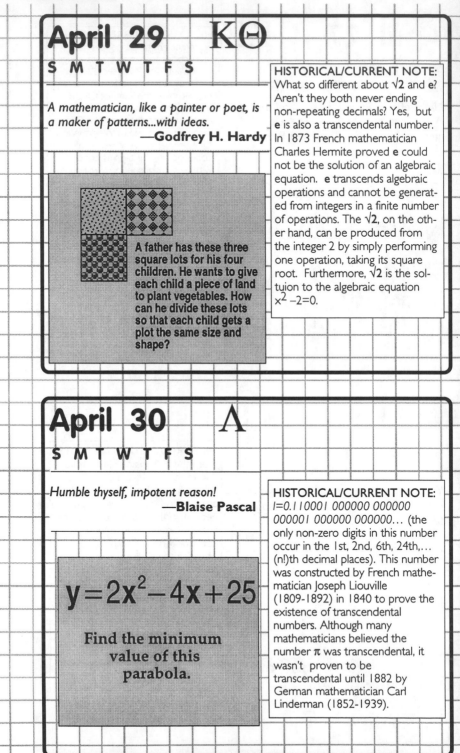

A father has these three square lots for his four children. He wants to give each child a piece of land to plant vegetables. How can he divide these lots so that each child gets a plot the same size and shape?

HISTORICAL/CURRENT NOTE:
What so different about √2 and **e**? Aren't they both never ending non-repeating decimals? Yes, but **e** is also a transcendental number. In 1873 French mathematician Charles Hermite proved **e** could not be the solution of an algebraic equation. **e** transcends algebraic operations and cannot be generated from integers in a finite number of operations. The √2, on the other hand, can be produced from the integer 2 by simply performing one operation, taking its square root. Furthermore, √2 is the soltuion to the algebraic equation $x^2 - 2 = 0$.

April 30 Λ

S M T W T F S

Humble thyself, impotent reason!
—**Blaise Pascal**

$$y = 2x^2 - 4x + 25$$

Find the minimum value of this parabola.

HISTORICAL/CURRENT NOTE:
l=0.110001 000000 000000 000001 000000 000000... (the only non-zero digits in this number occur in the 1st, 2nd, 6th, 24th,... (n!)th decimal places). This number was constructed by French mathematician Joseph Liouville (1809-1892) in 1840 to prove the existence of transcendental numbers. Although many mathematicians believed the number π was transcendental, it wasn't proven to be transcendental until 1882 by German mathematician Carl Linderman (1852-1939).

May 1   1

S M T W T F S

May dates are written in the binary number system

The infinite! no other question has moved so profoundly the Spirit of man
— **David Hilbert**

HISTORICAL/CURRENT NOTE: Accoring to Plutarch (c. 46-127 AD) Archimedes was completely consumed by mathematics: "... *being perpetually charmed by his familiar siren, that is, by his geometry, he neglected to eat and drink and took no care of his person; he was often carried by force to the baths, and when there he would trace geometrical figures in the ashes from the fire, and when it was anointed with oil he would draw lines with his finger upon his body, being in a state of great ecstasy and divinely possessed by his science.*"

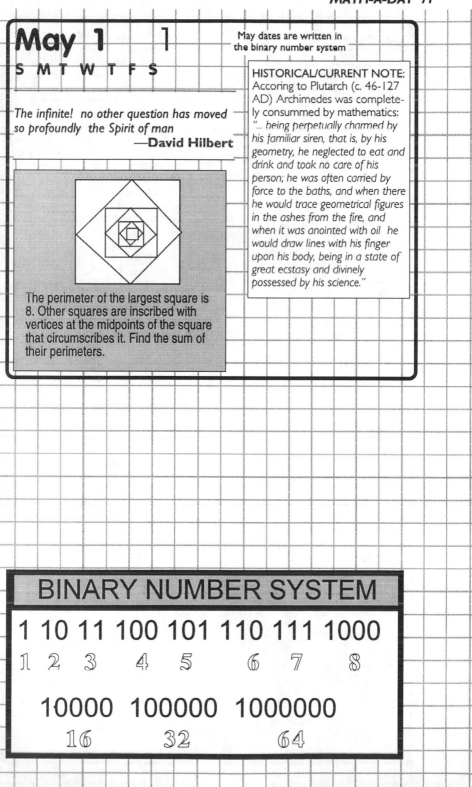

The perimeter of the largest square is 8. Other squares are inscribed with vertices at the midpoints of the square that circumscribes it. Find the sum of their perimeters.

BINARY NUMBER SYSTEM

1	10	11	100	101	110	111	1000
1	2	3	4	5	6	7	8

10000	100000	1000000
16	32	64

May 2   10

S M T W T F S

A monument to Newton! a monument to Shakespeare! Look up to Heaven, look into the Human Heart. Til the planets and the passions the affections and the fixed stars are extinguished their names cannot die. —**John Wilson**

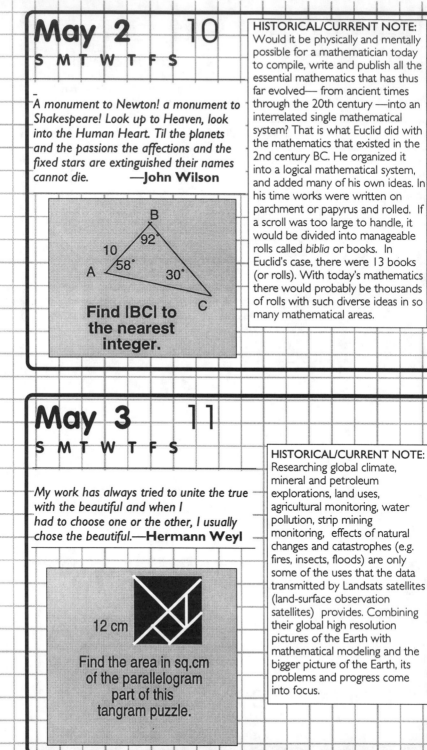

Find |BC| to the nearest integer.

HISTORICAL/CURRENT NOTE:
Would it be physically and mentally possible for a mathematician today to compile, write and publish all the essential mathematics that has thus far evolved— from ancient times through the 20th century —into an interrelated single mathematical system? That is what Euclid did with the mathematics that existed in the 2nd century BC. He organized it into a logical mathematical system, and added many of his own ideas. In his time works were written on parchment or papyrus and rolled. If a scroll was too large to handle, it would be divided into manageable rolls called *biblia* or books. In Euclid's case, there were 13 books (or rolls). With today's mathematics there would probably be thousands of rolls with such diverse ideas in so many mathematical areas.

May 3   11

S M T W T F S

My work has always tried to unite the true with the beautiful and when I had to choose one or the other, I usually chose the beautiful.—**Hermann Weyl**

12 cm

Find the area in sq.cm of the parallelogram part of this tangram puzzle.

HISTORICAL/CURRENT NOTE:
Researching global climate, mineral and petroleum explorations, land uses, agricultural monitoring, water pollution, strip mining monitoring, effects of natural changes and catastrophes (e.g. fires, insects, floods) are only some of the uses that the data transmitted by Landsats satellites (land-surface observation satellites) provides. Combining their global high resolution pictures of the Earth with mathematical modeling and the bigger picture of the Earth, its problems and progress come into focus.

May 4   100

S M T W T F S

From henceforth, space by itself, and time by itself, have vanished into the merest shadows and only a kind of blend of the two exists in its own right. **—Herman Minkowski**

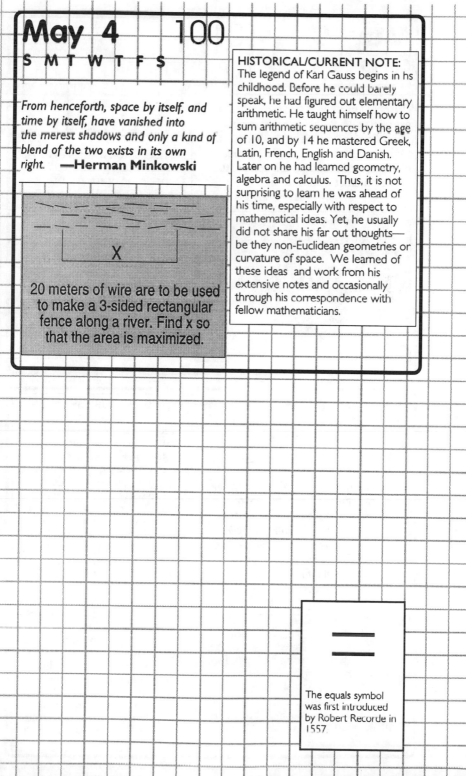

20 meters of wire are to be used to make a 3-sided rectangular fence along a river. Find x so that the area is maximized.

HISTORICAL/CURRENT NOTE:
The legend of Karl Gauss begins in his childhood. Before he could barely speak, he had figured out elementary arithmetic. He taught himself how to sum arithmetic sequences by the age of 10, and by 14 he mastered Greek, Latin, French, English and Danish. Later on he had learned geometry, algebra and calculus. Thus, it is not surprising to learn he was ahead of his time, especially with respect to mathematical ideas. Yet, he usually did not share his far out thoughts—be they non-Euclidean geometries or curvature of space. We learned of these ideas and work from his extensive notes and occasionally through his correspondence with fellow mathematicians.

The equals symbol was first introduced by Robert Recorde in 1557.

May 5   101

S M T W T F S

Ordinary language is totally unsuited for expressing what physics really asserts, since the words of everyday life are not sufficiently abstract. Only mathematics and mathematical logic can say as little as the physicist means to say..
— **Bertrand Russell**

HISTORICAL/CURRENT NOTE:
The first record of the use of a symbol for an operation dates back to ancient Egypt. The Ahmes Papyrus (c. 1550 BC) has symbols for addition ⊾ and subtraction ⟋ .

$$\frac{\left(\sqrt[100]{\text{googol}^2}\right)}{3^2}$$

May 6   110

S M T W T F S

Among the minor, yet striking characteristics of mathematics, may be mentioned the fleshless and skeletal build of its propositions; the peculiar difficulty, complication, and stress of its reasonings; the perfect exactitude of its results; their broad universality; their practical infallibility..
— **Charles Sanders Peirce**

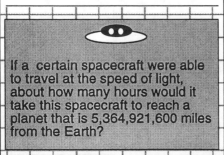
If a certain spacecraft were able to travel at the speed of light, about how many hours would it take this spacecraft to reach a planet that is 5,364,921,600 miles from the Earth?

HISTORICAL/CURRENT NOTE:
The first roles of mathematics involved counting and measuring. For measuring mass, the scientific world adopted the kilogram standard. A precisely machined cylinder made of platinum-iridium alloy is used as the standard weight for 1 kilogram. The cylinder is stored at the International Bureau of Weights and Measures in Sévres, France. The kilogram is the last of the standards that has such a physical form. Researchers are seeking a new standard for a kilogram based on an atomic mass such as that of a carbon atom. The National Institute of Standards and Technology in Maryland has weighed a kilogram against an electromagnetic force and is looking for methods to reach the required accuracy for the Planck constant. Another method being explored is counting the number of atoms in a kilogram of silicon. With such methods being perfected, the atomic kilogram is in the near future.

May 7   111

S M T W T F S

Religions die when they are proved true.
Science is the record of dead religions.
                              —Oscar Wilde

$$\sum_{n=1}^{\infty} 6\left(\frac{3}{4}\right)^{n-1}$$

HISTORICAL/CURRENT NOTE:
The °2nd millennium AD is the first millennium in the history of humanity whose major problems may lie in its chips— computer chips, that is. Any timing chip encoded with a year counter was potentially in jeopardy at the turn of the 20th century. Such chips are used in such varied places as industrial machinery, traffic control signals, monitoring devices, nuclear power plants, etc. In some cases the only way to solve the problem was to replace the chip — that is, if it was accessible.

May 8   1000

S M T W T F S

Mathematics is not a deductive science —
that's a cliche. When you try to prove a
theorem, you don't just list the hypotheses,
and then start to reason. What you do is
trial and error, experimentation, guess-
work.   **—Paul Halmos**

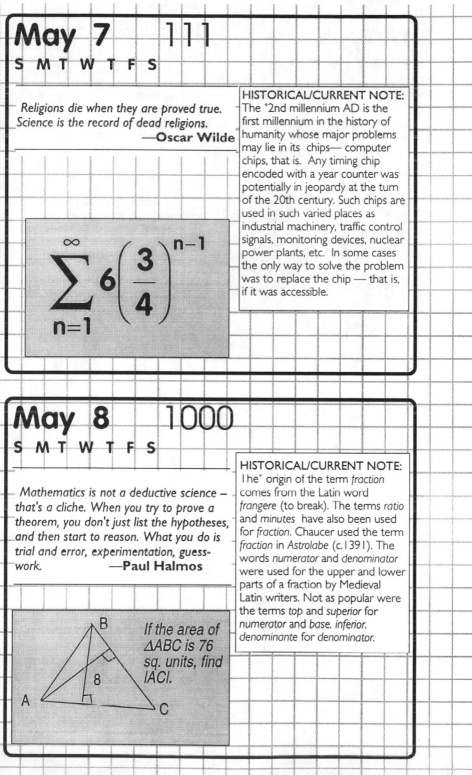

If the area of △ABC is 76 sq. units, find IACI.

HISTORICAL/CURRENT NOTE:
The° origin of the term *fraction* comes from the Latin word *frangere* (to break). The terms *ratio* and *minutes* have also been used for *fraction*. Chaucer used the term *fraction* in *Astrolabe* (c.1391). The words *numerator* and *denominator* were used for the upper and lower parts of a fraction by Medieval Latin writers. Not as popular were the terms *top* and *superior* for numerator and *base, inferior, denominante* for denominator.

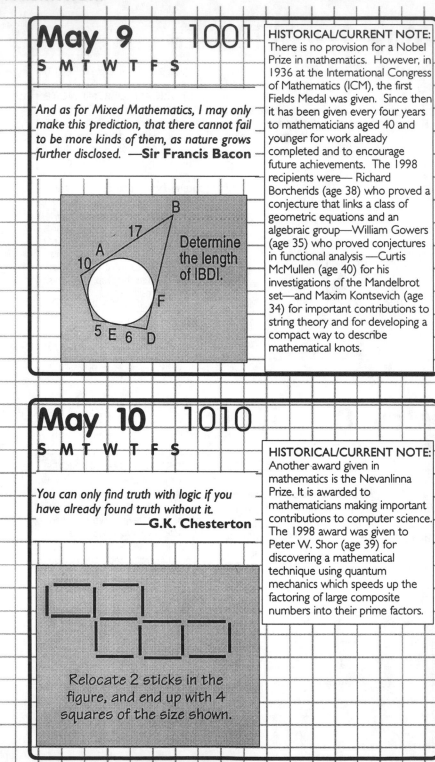

May 9   1001
S M T W T F S

And as for Mixed Mathematics, I may only make this prediction, that there cannot fail to be more kinds of them, as nature grows further disclosed. —**Sir Francis Bacon**

Determine the length of |BD|.

B
17
A
10
F
5 E 6 D

HISTORICAL/CURRENT NOTE: There is no provision for a Nobel Prize in mathematics. However, in 1936 at the International Congress of Mathematics (ICM), the first Fields Medal was given. Since then it has been given every four years to mathematicians aged 40 and younger for work already completed and to encourage future achievements. The 1998 recipients were— Richard Borcherids (age 38) who proved a conjecture that links a class of geometric equations and an algebraic group—William Gowers (age 35) who proved conjectures in functional analysis —Curtis McMullen (age 40) for his investigations of the Mandelbrot set—and Maxim Kontsevich (age 34) for important contributions to string theory and for developing a compact way to describe mathematical knots.

May 10   1010
S M T W T F S

You can only find truth with logic if you have already found truth without it.
—**G.K. Chesterton**

Relocate 2 sticks in the figure, and end up with 4 squares of the size shown.

HISTORICAL/CURRENT NOTE: Another award given in mathematics is the Nevanlinna Prize. It is awarded to mathematicians making important contributions to computer science. The 1998 award was given to Peter W. Shor (age 39) for discovering a mathematical technique using quantum mechanics which speeds up the factoring of large composite numbers into their prime factors.

May 11   1011

S **M T W T F** S

It may well be doubted whether, in all the range of science, there is any field so fascinating to the explorer—so rich in hidden treasures—so fruitful in delightful surprises—as Pure Mathematics.
—Lewis Carroll

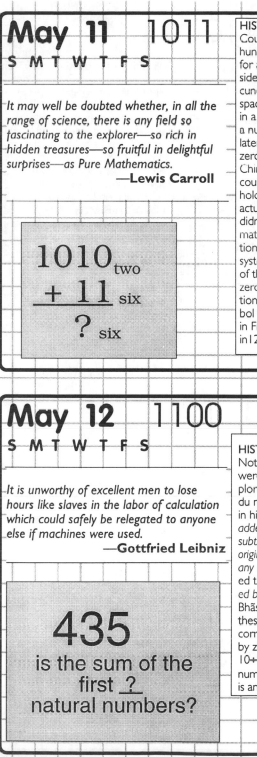

$$1010_{two}$$
$$+ 11_{six}$$
$$? _{six}$$

HISTORICAL/CURRENT NOTE:
Counting numbers were used for hundreds of years before the need for a zero ever arose or was considered. The Babylonians on their cuneiform clay tablets left a blank space for a zero placeholder within a number, but they never ended a number with such a space. They later developed a symbol for the zero placeholder. Similarly, the Chinese left a blank space on their counting boards for a zero placeholder. The first use of zero as an actual number and a placeholder didn't take place until the Indian mathematicians used 0 in conjunction with the base ten positional system they devised. The Mayans of the Yucatan also developed a zero placeholder with their positional number system. The 0 symbol was first introduced to Europe in Fibonacci's book *Liber Abaci* in 1202.

May 12   1100

S **M T W T F** S

It is unworthy of excellent men to lose hours like slaves in the labor of calculation which could safely be relegated to anyone else if machines were used.
—Gottfried Leibniz

435
is the sum of the first _?_ natural numbers?

HISTORICAL/CURRENT NOTE:
Not until around the 8th century were the properties of zero explored. First explored by the Hindu mathematician ´Sridhara, who in his work explained that *zero added to any number is zero, zero subtracted from any number is the original number, zero multiplied by any number is zero*—but he avoided the question of *a number divided by zero*. About 100 years later Bhāshara in *Lilàvati* pointed out these rules, and also specified (incorrectly) that a quantity divided by zero is zero. He illustrated that 10÷0=10/0, but also stated that a number with a zero denominator is an *indefinite term*.

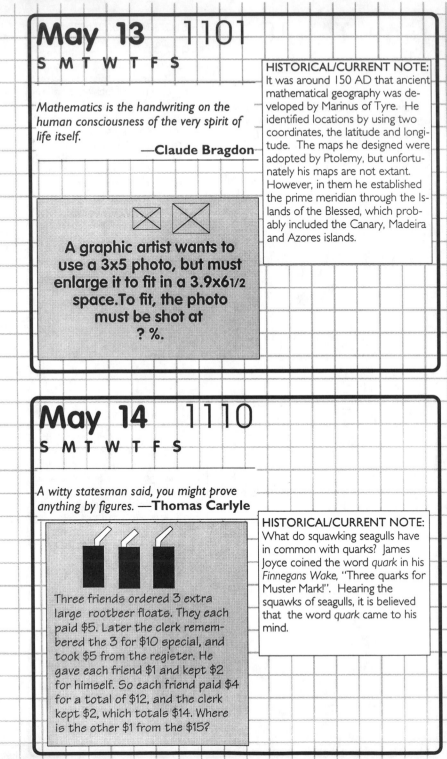

May 13   1101

S M T W T F S

Mathematics is the handwriting on the human consciousness of the very spirit of life itself.

—**Claude Bragdon**

A graphic artist wants to use a 3x5 photo, but must enlarge it to fit in a 3.9x61/2 space.To fit, the photo must be shot at ? %.

HISTORICAL/CURRENT NOTE:
It was around 150 AD that ancient mathematical geography was developed by Marinus of Tyre. He identified locations by using two coordinates, the latitude and longitude. The maps he designed were adopted by Ptolemy, but unfortunately his maps are not extant. However, in them he established the prime meridian through the Islands of the Blessed, which probably included the Canary, Madeira and Azores islands.

May 14   1110

S M T W T F S

A witty statesman said, you might prove anything by figures. —**Thomas Carlyle**

Three friends ordered 3 extra large rootbeer floats. They each paid $5. Later the clerk remembered the 3 for $10 special, and took $5 from the register. He gave each friend $1 and kept $2 for himself. So each friend paid $4 for a total of $12, and the clerk kept $2, which totals $14. Where is the other $1 from the $15?

HISTORICAL/CURRENT NOTE:
What do squawking seagulls have in common with quarks? James Joyce coined the word *quark* in his *Finnegans Wake*, ''Three quarks for Muster Mark!''. Hearing the squawks of seagulls, it is believed that the word *quark* came to his mind.

> The *greater than* and *less than* symbols we use today were first introduced by Thomas Harriot in 1631.

May 15

S M T W T F S

There is an astonishing imagination even in the science of mathematics.

—**Voltaire**

$$x+y=1$$
$$x-2z=5$$
$$2y+z=1$$
Find x, y, & z.

HISTORICAL/CURRENT NOTE:
The mathematics of converting money from one denomination to another is taking a dramatic new turn in Europe. The Euro is slowly taking over as a means of exchange among the countries of the European Common Market. In the past, some of these countries, such as Italy, was made up of independent states, each with its own monetary system. Thus, just making a short trip across Italy would involve a number of monetary exchanges. The global exchange system is undergoing a new twist, and the 21st century will see the growing popularity of *smart cards*. Rather than cash, credit cards or traveler checks, these cash cards will be immediately accepted for purchases without necessity of ID other than a PIN number.

May 16 10000

S M T W T F S

Men have become fools of their tools.
—Thoreau

A runner and a walker depart in opposite directions on a one mile track. They are going 9mph and 3mph respectively. In how many minutes will they meet?

HISTORICAL/CURRENT NOTE: How many people today have heard of Napier's work *Plaine Discovery of the Whole Revelation of Saint John?* Not many. Yet Scottish mathematician John Napier believed this was his most important work and contribution. Today, his invention of logarithms is his most valued contribution to the scientific world. The use of logarithms was invaluable in simplifying complex computations. He worked 20 years on his theory. Three years before he died, he published *Descripto,* a book of tables of logarithms. In 1617, *Rabdologia,* his book on Napier rods was published— the new computing devise of the 17th century captivated the interest of Europe and Asia alike.

May 17 10001

S M T W T F S

What is now proved was once only imagined. **—William Blake**

Area of △ADC=13
Find the area
of △ABC.

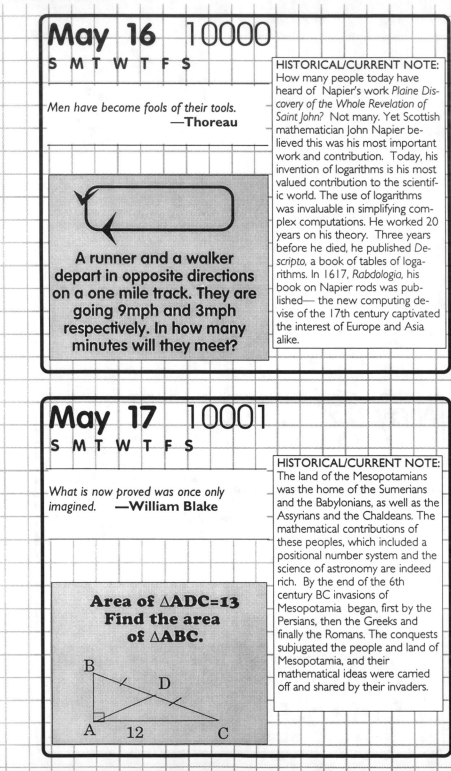

HISTORICAL/CURRENT NOTE: The land of the Mesopotamians was the home of the Sumerians and the Babylonians, as well as the Assyrians and the Chaldeans. The mathematical contributions of these peoples, which included a positional number system and the science of astronomy are indeed rich. By the end of the 6th century BC invasions of Mesopotamia began, first by the Persians, then the Greeks and finally the Romans. The conquests subjugated the people and land of Mesopotamia, and their mathematical ideas were carried off and shared by their invaders.

May 18 10010

S M T W T F S

Thus all human cognition begins with intuitions, proceeds from them to concepts and ends with ideas.

—Immanuel Kant

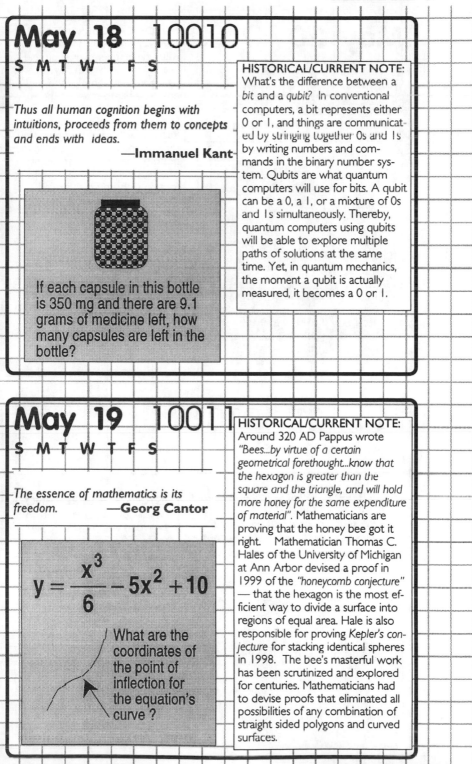

If each capsule in this bottle is 350 mg and there are 9.1 grams of medicine left, how many capsules are left in the bottle?

HISTORICAL/CURRENT NOTE:
What's the difference between a *bit* and a *qubit?* In conventional computers, a bit represents either 0 or 1, and things are communicated by stringing together 0s and 1s by writing numbers and commands in the binary number system. Qubits are what quantum computers will use for bits. A qubit can be a 0, a 1, or a mixture of 0s and 1s simultaneously. Thereby, quantum computers using qubits will be able to explore multiple paths of solutions at the same time. Yet, in quantum mechanics, the moment a qubit is actually measured, it becomes a 0 or 1.

May 19 10011

S M T W T F S

The essence of mathematics is its freedom.   **—Georg Cantor**

$$y = \frac{x^3}{6} - 5x^2 + 10$$

What are the coordinates of the point of inflection for the equation's curve?

HISTORICAL/CURRENT NOTE:
Around 320 AD Pappus wrote *"Bees...by virtue of a certain geometrical forethought...know that the hexagon is greater than the square and the triangle, and will hold more honey for the same expenditure of material"*. Mathematicians are proving that the honey bee got it right. Mathematician Thomas C. Hales of the University of Michigan at Ann Arbor devised a proof in 1999 of the *"honeycomb conjecture"* — that the hexagon is the most efficient way to divide a surface into regions of equal area. Hale is also responsible for proving *Kepler's conjecture* for stacking identical spheres in 1998. The bee's masterful work has been scrutinized and explored for centuries. Mathematicians had to devise proofs that eliminated all possibilities of any combination of straight sided polygons and curved surfaces.

May 20 10100
S M T W T F S

There is no nature at an instant.
—Alfred North Whitehead

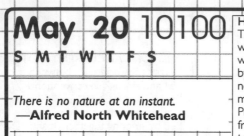

$$AAB$$
$$\times \quad B$$
$$\overline{CB5B}$$

If A, B, and C are all
different digits,
which digits are they?

HISTORICAL/CURRENT NOTE:
The way we write fractions today, with one number over another, was a notation system developed by the Hindus, although they did not use a bar to separate the numerator from the denominator. Prior to this, the Egyptians wrote fractions with an implied numerator of I and a denominator on the bottom. For example, ⵚ meant 1/2, with the symbol ⵔ meaning a fraction with a numerator of I. The use of the bar was introduced by Arab writers, and in the Middle Ages the bar appeared in Latin manuscripts. Writing fractions with the bar presented problems in early printed works, especially with setting the type. Often, for convenience, it was necessary to leave off the bar or write fractions in other ways, for example,

$$3|4, \quad \frac{3}{4}, \quad \text{and} \quad \frac{2}{3\ 5}$$

May 21 10101
S M T W T F S

There is music even in the beauty and the silent note which Cupid strikes, far sweeter than the sound of an instrument; for there is music wherever there is harmony, order, or proportion; and thus far we may maintain the music of the spheres.
—Sir Thomas Browne

HISTORICAL/CURRENT NOTE:
When Columbus set out in 1492, the world was considered flat. Yet in 450 BC, Herodotus wrote about Parmenides of Elea (circa 460 BC) who believed the Earth was spherical. Also in Herodotus' historical account appeared the first reference to the concept of a meridian.

$$\frac{0.01}{x} = \frac{5}{11000}$$

May 22 10110
S M T W T F S

It is easier to square the circle than to get round a mathematician.
—**Augustus De Morgan**

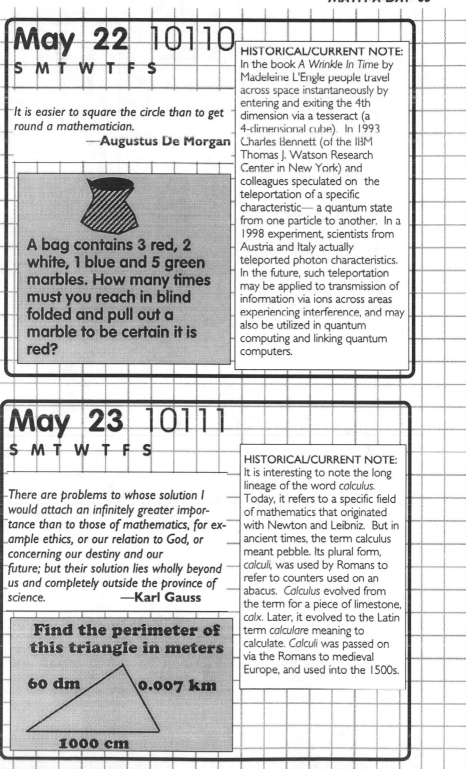

A bag contains 3 red, 2 white, 1 blue and 5 green marbles. How many times must you reach in blind folded and pull out a marble to be certain it is red?

HISTORICAL/CURRENT NOTE:
In the book *A Wrinkle In Time* by Madeleine L'Engle people travel across space instantaneously by entering and exiting the 4th dimension via a tesseract (a 4-dimensional cube). In 1993 Charles Bennett (of the IBM Thomas J. Watson Research Center in New York) and colleagues speculated on the teleportation of a specific characteristic— a quantum state from one particle to another. In a 1998 experiment, scientists from Austria and Italy actually teleported photon characteristics. In the future, such teleportation may be applied to transmission of information via ions across areas experiencing interference, and may also be utilized in quantum computing and linking quantum computers.

May 23 10111
S M T W T F S

There are problems to whose solution I would attach an infinitely greater importance than to those of mathematics, for example ethics, or our relation to God, or concerning our destiny and our future; but their solution lies wholly beyond us and completely outside the province of science.   —**Karl Gauss**

HISTORICAL/CURRENT NOTE:
It is interesting to note the long lineage of the word *calculus*. Today, it refers to a specific field of mathematics that originated with Newton and Leibniz. But in ancient times, the term calculus meant pebble. Its plural form, *calculi*, was used by Romans to refer to counters used on an abacus. *Calculus* evolved from the term for a piece of limestone, *calx*. Later, it evolved to the Latin term *calculare* meaning to calculate. *Calculi* was passed on via the Romans to medieval Europe, and used into the 1500s.

Find the perimeter of this triangle in meters

60 dm   **0.007 km**

1000 cm

Π

The Greek capital letter, pi, is the symbol René Descartes used for *product* in the 17th century.

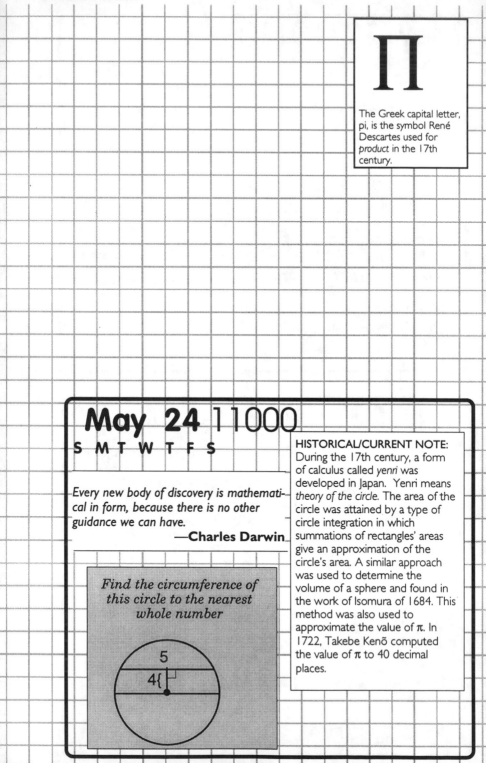

May 24 11000

S M T W T F S

Every new body of discovery is mathematical in form, because there is no other guidance we can have.

—Charles Darwin

Find the circumference of this circle to the nearest whole number

5

4{

HISTORICAL/CURRENT NOTE: During the 17th century, a form of calculus called *yenri* was developed in Japan. Yenri means *theory of the circle.* The area of the circle was attained by a type of circle integration in which summations of rectangles' areas give an approximation of the circle's area. A similar approach was used to determine the volume of a sphere and found in the work of Isomura of 1684. This method was also used to approximate the value of π. In 1722, Takebe Kenō computed the value of π to 40 decimal places.

May 25 11001

S M T W T F S

Science is always wrong; it never solves a problem without creating ten more.
—George Bernard Shaw

A tennis ball is dropped at a point 16" from the ground. It continually rebounds 3/7th's of its previous height. How many inches will it have bounced before coming to rest?

HISTORICAL/CURRENT NOTE:
In the future conducting of experiments will rely more and more on virtual reality. The virtual laboratory and virtual reaction will eventually become the norm, and the virtual microscope, when perfected, will be an invaluable tool which will allow invisible chemical behavior of molecules to be seen. In addition, experimentation with varied conditions (e.g. changes in temperatures and pressures) will be easily performed — experiments will be viewed backwards and forwards by playing with the time scale — visual perspective of a reaction will be easily changed — and simulations will be far safer than the real thing. The molecular world relies on quantum mechanics and Schrodinger's equation. Chemical reactions occur on the scale of picoseconds (1 picosecond=1 trillionth of a second). When will all this happen? As soon as supercomputer hardware can handle the enormous amount of data required to describe such experiments. A test model is in the works with NEC's latest computer simulation technology.

May 26 11010

S M T W T F S

To see the World in a grain of sand, And a Heaven in a wildflower, Hold Infinity in the palm of your hand, And Eternity in an hour.
—William Blake

HISTORICAL/CURRENT NOTE:
It was difficult to use Roman numerals for computing, which may account for why Romans relied on the abacus for computation. They used three different types of abaci. The earliest was the *dust abacus* in which a table or board was covered with sand or fine dust, and marks were made with a stylus and erased by hand. In essence, it was an ancient type of etch-n-sketch. Another type used beads in grooves, and the third was made to be used with counters (called calculi or abaculi). The boards (known as tables) were either made from wax, marble, wood or metal., and the counters were either pebbles, ivory or glass.

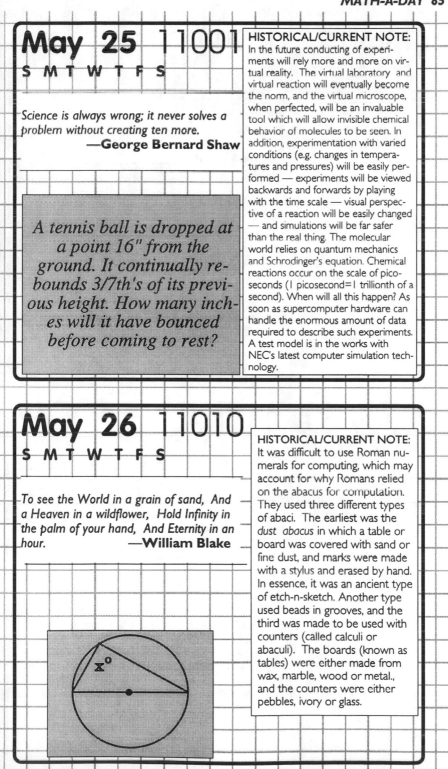

May 27 11011

S M T W T F S

If I feel unhappy, I do mathematics to become happy. If I am happy, I do mathematics to keep happy..
—**Alfréd Rényi**

HISTORICAL/CURRENT NOTE:
The word *abacus* comes from the Greek word $\alpha\beta\alpha\sigma\iota\varsigma$ pronounced *abasis* and meaning *without base*. This described the table that had no legs on which computation was performed.

$$81^{\frac{x}{2}-4} = \frac{1}{9^{-1}}$$

May 28 11100

S M T W T F S

Mathematics is the supreme arbiter. From its decisions there is no appeal.
—**Tobias Dantzig**

HISTORICAL/CURRENT NOTE:
The 20th century introduced the *Theory of Everything* (aka TOE or String Theory) in the 1960s. The theory explains that the essence of the universe lies not with atoms, molecules, muons, photons, gluons, quarks, etc., but with strings which reside in multiple dimensions. The vibrations of these strings are what defines the various forms of matter. *String Theory* reconciles Einstein's *General Theory of Relativity* and *quantum mechanics*, which is a phenomenal marriage of mathematics and physics.

$A = \{\text{naturals} < 50\}$

$B = \{\text{multiples of } 5\}$

$C = \{\text{odd perfect sqs.}\}$

$A \cap B \cap C = ?$

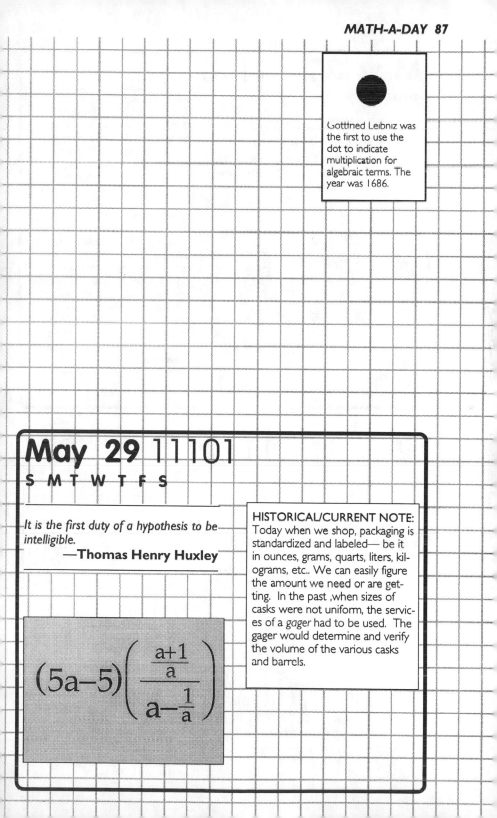

Gottfried Leibniz was the first to use the dot to indicate multiplication for algebraic terms. The year was 1686.

May 29 11101
S M T W T F S

It is the first duty of a hypothesis to be intelligible.

—Thomas Henry Huxley

$$(5a-5)\left(\dfrac{\dfrac{a+1}{a}}{a-\dfrac{1}{a}}\right)$$

HISTORICAL/CURRENT NOTE:
Today when we shop, packaging is standardized and labeled— be it in ounces, grams, quarts, liters, kilograms, etc.. We can easily figure the amount we need or are getting. In the past ,when sizes of casks were not uniform, the services of a *gager* had to be used. The gager would determine and verify the volume of the various casks and barrels.

May 30 11110

S M T W T F S

The mathematician's patterns, like the painter's or the poet's must be beautiful; the ideas, like the colors or the words must fit together in a harmonious way.
—Godfrey Hardy

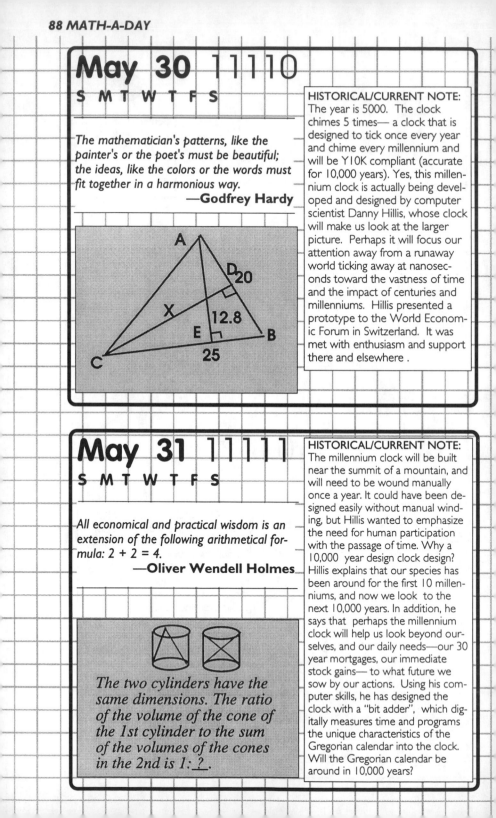

HISTORICAL/CURRENT NOTE: The year is 5000. The clock chimes 5 times— a clock that is designed to tick once every year and chime every millennium and will be Y10K compliant (accurate for 10,000 years). Yes, this millennium clock is actually being developed and designed by computer scientist Danny Hillis, whose clock will make us look at the larger picture. Perhaps it will focus our attention away from a runaway world ticking away at nanoseconds toward the vastness of time and the impact of centuries and millenniums. Hillis presented a prototype to the World Economic Forum in Switzerland. It was met with enthusiasm and support there and elsewhere .

May 31 11111

S M T W T F S

All economical and practical wisdom is an extension of the following arithmetical formula: 2 + 2 = 4.
—Oliver Wendell Holmes

The two cylinders have the same dimensions. The ratio of the volume of the cone of the 1st cylinder to the sum of the volumes of the cones in the 2nd is 1: ? .

HISTORICAL/CURRENT NOTE: The millennium clock will be built near the summit of a mountain, and will need to be wound manually once a year. It could have been designed easily without manual winding, but Hillis wanted to emphasize the need for human participation with the passage of time. Why a 10,000 year design clock design? Hillis explains that our species has been around for the first 10 millenniums, and now we look to the next 10,000 years. In addition, he says that perhaps the millennium clock will help us look beyond ourselves, and our daily needs—our 30 year mortgages, our immediate stock gains— to what future we sow by our actions. Using his computer skills, he has designed the clock with a "bit adder", which digitally measures time and programs the unique characteristics of the Gregorian calendar into the clock. Will the Gregorian calendar be around in 10,000 years?

June 1

S M T W T F S

Nobody knew then that there could be space. ...in reality there wasn't even space to pack us into. Every point of each of us coincided with every point of each of the others in a single point, which is where we all were.
—Italo Calvino,
All In One Point

June dates are written in the the Egyptian hieratic number system

HISTORICAL/CURRENT NOTE:
One of the most famous problems of the Middle Ages was the *chessboard & the grains* problem. One Arab story tells us that King Shihram was so delighted with the game of chess invented by Sissah ibn Dah that the king insisted Sissah choose any reward he wished. Sissah only asked for grain, specifying the amount by using his chessboard. *One grain of wheat for the 1st square of the chessboard, 2 for the 2nd, 4 for the 3rd...2^{64} for the last square.* This ended up being more grain than the kingdom could produce. The problem dates back at least to the 10th century, and traveled to Europe via Fibonacci's book *Liber Abaci* of 1202. It appeared in many manuscripts of the 13th, 14th, and 15th centuries, where the story was changed to suit the time and region.

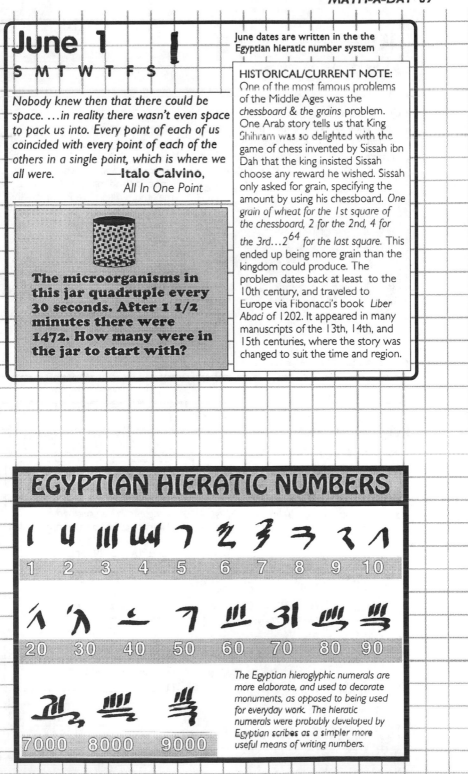

The microorganisms in this jar quadruple every 30 seconds. After 1 1/2 minutes there were 1472. How many were in the jar to start with?

EGYPTIAN HIERATIC NUMBERS

ı	Ч	ııı	Ш	7	ʑ	ʒ	ꝯ	ʓ	ʌ
1	2	3	4	5	6	7	8	9	10

ʌ	ʌ	ᴗ	7	Ш	3ı	Ш	Ш
20	30	40	50	60	70	80	90

Ш	Ш	Ш
7000	8000	9000

The Egyptian hieroglyphic numerals are more elaborate, and used to decorate monuments, as opposed to being used for everyday work. The hieratic numerals were probably developed by Egyptian scribes as a simpler more useful means of writing numbers.

June 2   4
S M T W T F S

Quaternions came from Hamilton after his really good work had been done; and though beautifully ingenious, have been an unmixed evil to those who have touched them in any way. —**Lord Kelvin**

$$10^? = \text{googolplex}$$

HISTORICAL/CURRENT NOTE:
The Great Pyramid of Giza (built between 2600 to 2500 BC) was an amazing feat of architecture, engineering and mathematics. The maximum difference in the lengths of its sides was 63/100 of an inch and its corners are within 1/27000 of a 90° angle. In addition, it is interesting to note that the perimeter of its base divided by its height is 2π, when π is taken as 3.14.

June 3   ///
S M T W T F S

Why, sometimes I've believed as many as six impossible things before breakfast . —**Lewis Carroll**

The snail problem
Suppose a snail is climbing up a slippery 30-inch wall. Each minute it climbs 5 inches, but slides back 4 inches. How many minutes will it take the snail to reach the top of the wall?

HISTORICAL/CURRENT NOTE:
What is known as the *table of chords (of double sines of half the angle)* can be considered the first introduction to trigonometry. It was developed by Hipparchus (c.190-c.126 BC) on the island of Rhodes. Hipparchus is also famous for his impressive astronomical work in which he measured angles and distances on a sphere, thus introducing spherical trigonometry. He also represented the projection of a celestial sphere on a plane, and catalogued 850 stars.

June 4
S M T W T F S ⩊

It is the perennial youthfulness of mathematics itself which marks it off with a disconcerting immorality from the other sciences. —**Eric Temple Bell**

$$\frac{9}{10} + \frac{9}{100} + \frac{9}{1000} + \cdots$$

$$= ?$$

n={natural numbers}

HISTORICAL/CURRENT NOTE:
Virtual reality (worlds created in cyberspace using computer technologies and sophisticated gear) has been used in many scientific experiments. For example, combining *functional magnetic resonance imaging (fMRI)* has been combined with the virtual reality used in computer games involving mazes. Humans are put through a variety of maze experiments while the fMRI is used to view which parts of the brain are activated in this type of problem solving. Such experimentation may someday help unlock some of the secrets of brain

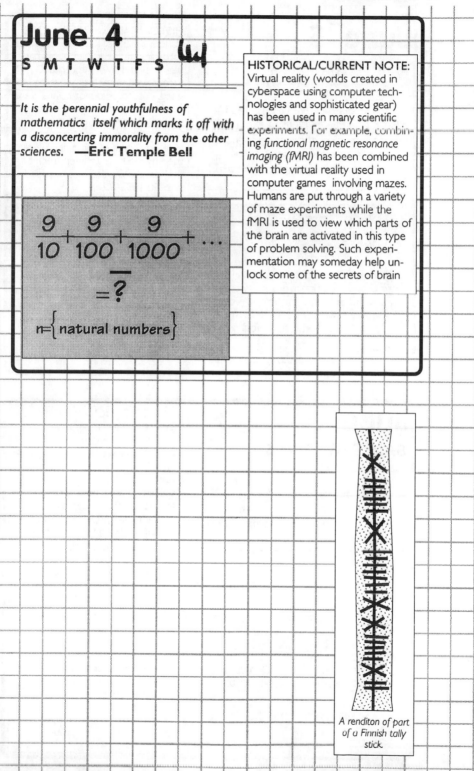

A renditon of part of a Finnish tally stick.

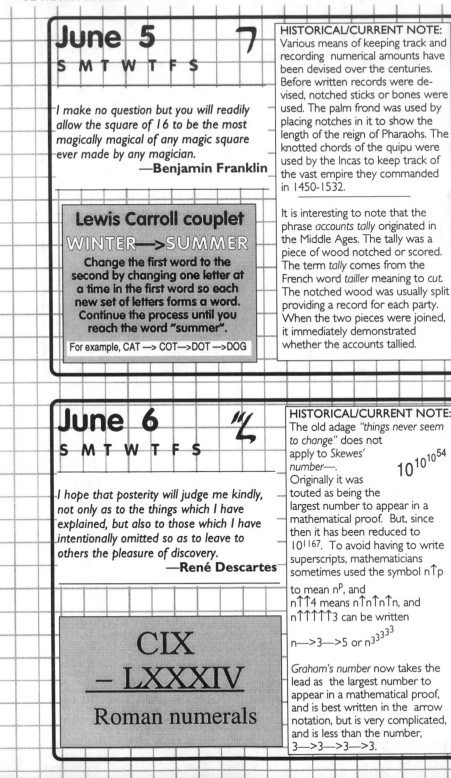

June 5

S M T W T F S

7

I make no question but you will readily allow the square of 16 to be the most magically magical of any magic square ever made by any magician.

—**Benjamin Franklin**

Lewis Carroll couplet

WINTER —> SUMMER

Change the first word to the second by changing one letter at a time in the first word so each new set of letters forms a word. Continue the process until you reach the word "summer".

For example, CAT —> COT —> DOT —> DOG

HISTORICAL/CURRENT NOTE:
Various means of keeping track and recording numerical amounts have been devised over the centuries. Before written records were devised, notched sticks or bones were used. The palm frond was used by placing notches in it to show the length of the reign of Pharaohs. The knotted chords of the quipu were used by the Incas to keep track of the vast empire they commanded in 1450-1532.

It is interesting to note that the phrase *accounts tally* originated in the Middle Ages. The tally was a piece of wood notched or scored. The term *tally* comes from the French word *tailler* meaning to *cut.* The notched wood was usually split providing a record for each party. When the two pieces were joined, it immediately demonstrated whether the accounts tallied.

June 6

S M T W T F S

I hope that posterity will judge me kindly, not only as to the things which I have explained, but also to those which I have intentionally omitted so as to leave to others the pleasure of discovery.

—**René Descartes**

CIX
− LXXXIV

Roman numerals

HISTORICAL/CURRENT NOTE:
The old adage *"things never seem to change"* does not apply to *Skewes' number*—.

$$10^{10^{10^{54}}}$$

Originally it was touted as being the largest number to appear in a mathematical proof. But, since then it has been reduced to 10^{1167}. To avoid having to write superscripts, mathematicians sometimes used the symbol $n{\uparrow}p$ to mean n^p, and $n{\uparrow}{\uparrow}4$ means $n{\uparrow}n{\uparrow}n{\uparrow}n$, and $n{\uparrow}{\uparrow}{\uparrow}{\uparrow}{\uparrow}3$ can be written

n—>3—>5 or $n^{3^{3^{3^{3^3}}}}$

Graham's number now takes the lead as the largest number to appear in a mathematical proof, and is best written in the arrow notation, but is very complicated, and is less than the number, 3—>3—>3—>3.

June 7

S M T W T F S

...men do not sufficiently understand the excellent use of the pure mathematics...—**Roger Bacon**

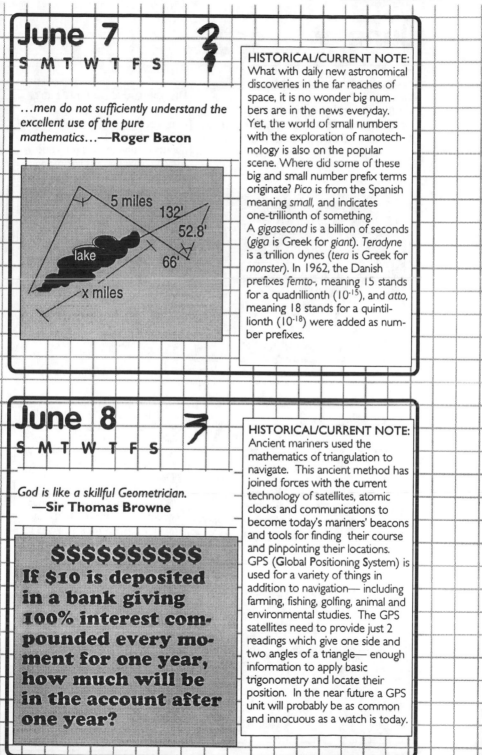

5 miles

132'

52.8'

lake

66'

x miles

HISTORICAL/CURRENT NOTE:
What with daily new astronomical discoveries in the far reaches of space, it is no wonder big numbers are in the news everyday. Yet, the world of small numbers with the exploration of nanotechnology is also on the popular scene. Where did some of these big and small number prefix terms originate? *Pico* is from the Spanish meaning *small*, and indicates one-trillionth of something. A *gigasecond* is a billion of seconds (*giga* is Greek for *giant*). *Teradyne* is a trillion dynes (*tera* is Greek for *monster*). In 1962, the Danish prefixes *femto-*, meaning 15 stands for a quadrillionth (10^{-15}), and *atto*, meaning 18 stands for a quintillionth (10^{-18}) were added as number prefixes.

June 8

S M T W T F S

God is like a skillful Geometrician.
—**Sir Thomas Browne**

**$$$$$$$$$$
If $10 is deposited
in a bank giving
100% interest compounded every moment for one year,
how much will be
in the account after
one year?**

HISTORICAL/CURRENT NOTE:
Ancient mariners used the mathematics of triangulation to navigate. This ancient method has joined forces with the current technology of satellites, atomic clocks and communications to become today's mariners' beacons and tools for finding their course and pinpointing their locations. GPS (Global Positioning System) is used for a variety of things in addition to navigation— including farming, fishing, golfing, animal and environmental studies. The GPS satellites need to provide just 2 readings which give one side and two angles of a triangle— enough information to apply basic trigonometry and locate their position. In the near future a GPS unit will probably be as common and innocuous as a watch is today.

June 9
S M T W T F S

Each problem that I solved became a rule which served afterwards to solve other problems. —**René Descartes**

For this expression to be real,

$$\sqrt{39 - 3x}$$

x must be ≤ to __?__.

HISTORICAL/CURRENT NOTE:
Today π and its infinite non-repeating decimals still intrigue mathematicians. The decimal approximation for π is now out millions of places. It is fascinating to note that in 1873, after 15 years of computing, William Shanks claimed to have calculated π to 707 places. His record held until 1949, when the ENIAK computer calculated π to 2035 places, and found an error in Shanks computation after the 500 and so digit.

June 10
S M T W T F S

Thus metaphysics and mathematics are, among all the sciences that belong to reason, those in which imagination has the greatest role. —**Jean D'Alembert**

$$n\overline{)721} \quad 37 \text{ r}18$$

HISTORICAL/CURRENT NOTE:
The U.S. Constitution provided for a count of population *"within three years after the first meeting of the Congress of the U.S., and within every subsequent term of 10 years."*

The first U.S. census was taken in 1790. The censuses were manually counted until the 1890 census. The 1880 census took 10 years to count. By the time it was finished, it was time for the next census. By necessity, a faster method had to be devised. A competition was held by U.S. Census Department for a more efficient method. Herman Hollerith designed a machine that processed information from punched cards using counter wheels and electromagnetic relays. His machine won the trial run, and later processed the 1890 census in just one month! This was one of many innovations that laid the ground work for today's computers.

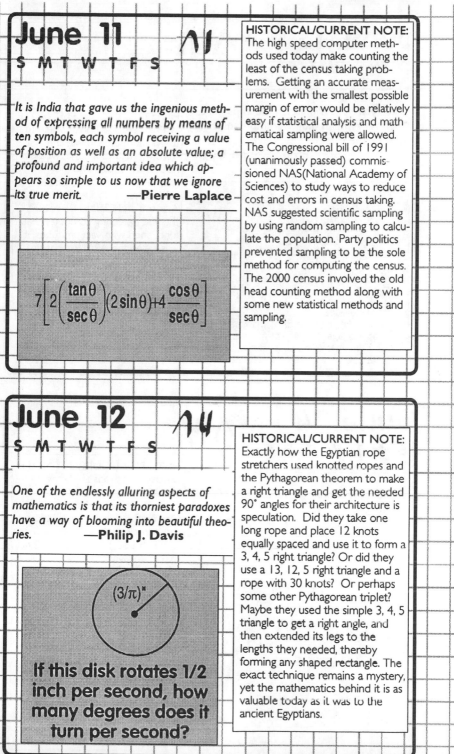

June 11

S M T W T F S

It is India that gave us the ingenious method of expressing all numbers by means of ten symbols, each symbol receiving a value of position as well as an absolute value; a profound and important idea which appears so simple to us now that we ignore its true merit. —**Pierre Laplace**

$$7\left[2\left(\frac{\tan\theta}{\sec\theta}\right)(2\sin\theta)+4\frac{\cos\theta}{\sec\theta}\right]$$

HISTORICAL/CURRENT NOTE:
The high speed computer methods used today make counting the least of the census taking problems. Getting an accurate measurement with the smallest possible margin of error would be relatively easy if statistical analysis and mathematical sampling were allowed. The Congressional bill of 1991 (unanimously passed) commissioned NAS(National Academy of Sciences) to study ways to reduce cost and errors in census taking. NAS suggested scientific sampling by using random sampling to calculate the population. Party politics prevented sampling to be the sole method for computing the census. The 2000 census involved the old head counting method along with some new statistical methods and sampling.

June 12

S M T W T F S

One of the endlessly alluring aspects of mathematics is that its thorniest paradoxes have a way of blooming into beautiful theories. —**Philip J. Davis**

$(3/\pi)^{\ast}$

If this disk rotates 1/2 inch per second, how many degrees does it turn per second?

HISTORICAL/CURRENT NOTE:
Exactly how the Egyptian rope stretchers used knotted ropes and the Pythagorean theorem to make a right triangle and get the needed 90° angles for their architecture is speculation. Did they take one long rope and place 12 knots equally spaced and use it to form a 3, 4, 5 right triangle? Or did they use a 13, 12, 5 right triangle and a rope with 30 knots? Or perhaps some other Pythagorean triplet? Maybe they used the simple 3, 4, 5 triangle to get a right angle, and then extended its legs to the lengths they needed, thereby forming any shaped rectangle. The exact technique remains a mystery, yet the mathematics behind it is as valuable today as it was to the ancient Egyptians.

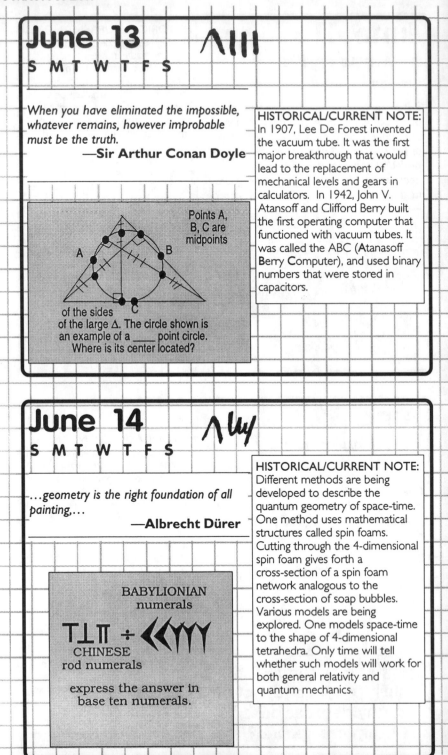

June 13
S M T W T F S

Λ|||

When you have eliminated the impossible, whatever remains, however improbable must be the truth.

—**Sir Arthur Conan Doyle**

Points A, B, C are midpoints

A   B

of the sides C of the large △. The circle shown is an example of a _____ point circle. Where is its center located?

HISTORICAL/CURRENT NOTE:
In 1907, Lee De Forest invented the vacuum tube. It was the first major breakthrough that would lead to the replacement of mechanical levels and gears in calculators. In 1942, John V. Atansoff and Clifford Berry built the first operating computer that functioned with vacuum tubes. It was called the ABC (**A**tanasoff **B**erry **C**omputer), and used binary numbers that were stored in capacitors.

June 14
S M T W T F S

Λ|Ʉɣ

...geometry is the right foundation of all painting,...

—**Albrecht Dürer**

BABYLIONIAN numerals

T⊥π ÷ ⟪⟪ᛉᛉ

CHINESE rod numerals

express the answer in base ten numerals.

HISTORICAL/CURRENT NOTE:
Different methods are being developed to describe the quantum geometry of space-time. One method uses mathematical structures called spin foams. Cutting through the 4-dimensional spin foam gives forth a cross-section of a spin foam network analogous to the cross-section of soap bubbles. Various models are being explored. One models space-time to the shape of 4-dimensional tetrahedra. Only time will tell whether such models will work for both general relativity and quantum mechanics.

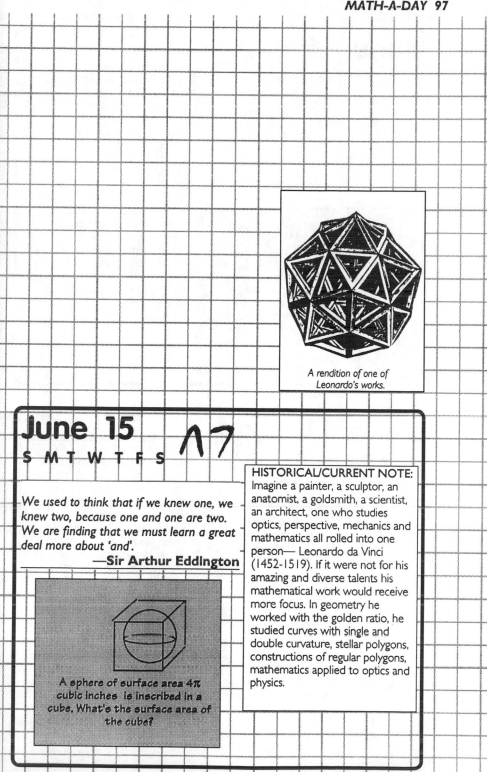

A rendition of one of
Leonardo's works.

June 15
S M T W T F S ∧7

*We used to think that if we knew one, we
knew two, because one and one are two.
We are finding that we must learn a great
deal more about 'and'.*

—**Sir Arthur Eddington**

A sphere of surface area 4π
cubic inches is inscribed in a
cube. What's the surface area of
the cube?

HISTORICAL/CURRENT NOTE:
Imagine a painter, a sculptor, an
anatomist, a goldsmith, a scientist,
an architect, one who studies
optics, perspective, mechanics and
mathematics all rolled into one
person— Leonardo da Vinci
(1452-1519). If it were not for his
amazing and diverse talents his
mathematical work would receive
more focus. In geometry he
worked with the golden ratio, he
studied curves with single and
double curvature, stellar polygons,
constructions of regular polygons,
mathematics applied to optics and
physics.

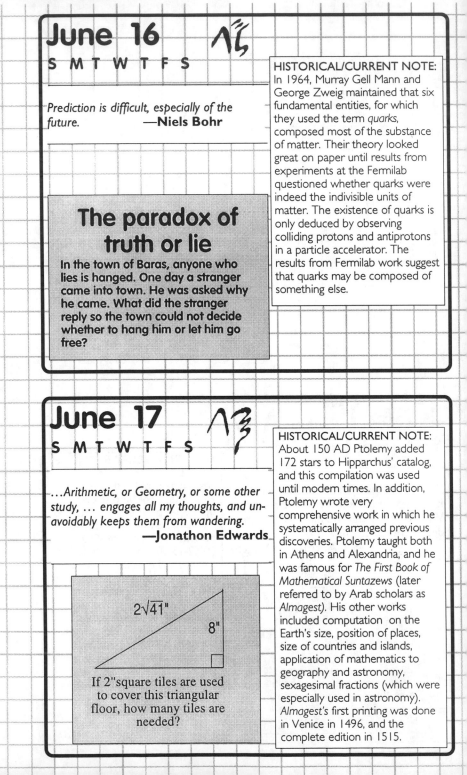

June 16
S M T W T F S

Prediction is difficult, especially of the future. —**Niels Bohr**

The paradox of truth or lie

In the town of Baras, anyone who lies is hanged. One day a stranger came into town. He was asked why he came. What did the stranger reply so the town could not decide whether to hang him or let him go free?

HISTORICAL/CURRENT NOTE:
In 1964, Murray Gell Mann and George Zweig maintained that six fundamental entities, for which they used the term *quarks*, composed most of the substance of matter. Their theory looked great on paper until results from experiments at the Fermilab questioned whether quarks were indeed the indivisible units of matter. The existence of quarks is only deduced by observing colliding protons and antiprotons in a particle accelerator. The results from Fermilab work suggest that quarks may be composed of something else.

June 17
S M T W T F S

...Arithmetic, or Geometry, or some other study, ... engages all my thoughts, and unavoidably keeps them from wandering. —**Jonathon Edwards**

$2\sqrt{41}"$

$8"$

If 2"square tiles are used to cover this triangular floor, how many tiles are needed?

HISTORICAL/CURRENT NOTE:
About 150 AD Ptolemy added 172 stars to Hipparchus' catalog, and this compilation was used until modern times. In addition, Ptolemy wrote very comprehensive work in which he systematically arranged previous discoveries. Ptolemy taught both in Athens and Alexandria, and he was famous for *The First Book of Mathematical Suntazews* (later referred to by Arab scholars as *Almagest*). His other works included computation on the Earth's size, position of places, size of countries and islands, application of mathematics to geography and astronomy, sexagesimal fractions (which were especially used in astronomy). *Almagest's* first printing was done in Venice in 1496, and the complete edition in 1515.

June 18

S M T W T F S

Mathematics is not only real, but it is the only reality. ...

—**Martin Gardner**

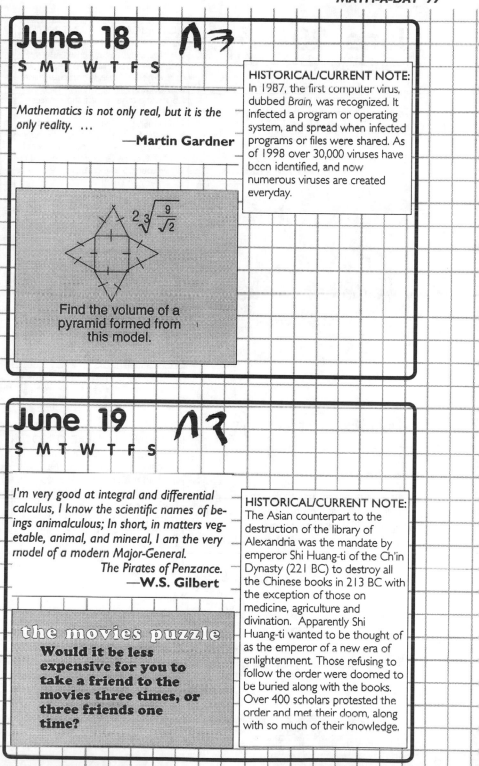

$$2\sqrt[3]{\frac{9}{\sqrt{2}}}$$

Find the volume of a pyramid formed from this model.

June 19

S M T W T F S

I'm very good at integral and differential calculus, I know the scientific names of beings animalculous; In short, in matters vegetable, animal, and mineral, I am the very model of a modern Major-General.
The Pirates of Penzance.
—**W.S. Gilbert**

the movies puzzle

Would it be less expensive for you to take a friend to the movies three times, or three friends one time?

June 20

S M T W T F S

To be a scholar of mathematics you must be born with talent, insight, concentration, taste, luck, drive and the ability to visualize and guess.   **—Paul Halmos**

Find x so that—

$$8x^3 - 1728 = 0$$

and x is a real number.

HISTORICAL/CURRENT NOTE:
During ancient times, the Dark Ages, the Middle Ages and the Renaissance, merchants, tradesmen, pilgrims, and armies exported ideas, in particular mathematical ideas, along with spices, teas, fabrics, and other goods and services. This is how the Hindu-Arabic numerals and numeration eventually were adopted by the Europeans. Along with the invasion of India by Alexander the Great in 327 BC came Greeks who resided in Indian courts, and exchanged ideas of astronomy and mathematics with the Hindu. In the 7th century a Sanskrit calendar was translated into Chinese. In 618, a Hindu astronomer was hired to devise a new calendar for the Chinese Bureau of Astronomy. The Romans sent an ambassador to the Chinese court in 719. In the 7th century, Bagdad was the center of the mathematical world. Arab geographer Mas'^udi mentions India, Ceylon, and China in his book *Meadows of Gold* of 915. The list of exchanges goes on and on, as it does today with ideas exported and imported throughout the world via books, TV, the Internet, periodicals.

June 21

S M T W T F S

The real danger is not that computers will begin to think like men, but that men will begin to think like computers.
                    —Sidney J. Harris

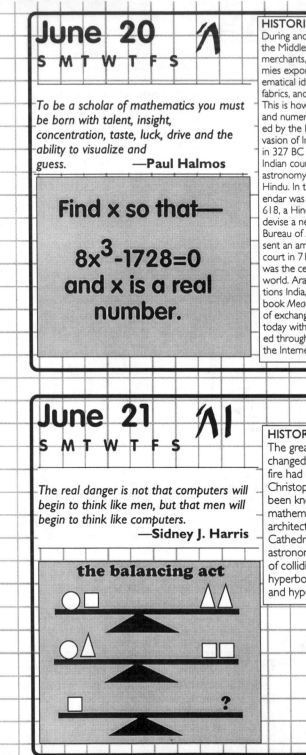

the balancing act

HISTORICAL/CURRENT NOTE:
The great fire of London in 1666 changed many people's lives. If the fire had not occurred, perhaps Christopher Wren would have been known today for his mathematics rather than his architectural work on St. Paul's Cathedral. His works included astronomy, the study of the nature of colliding bodies, the grinding of hyperbolic mirrors, perspective, and hyperboloids.

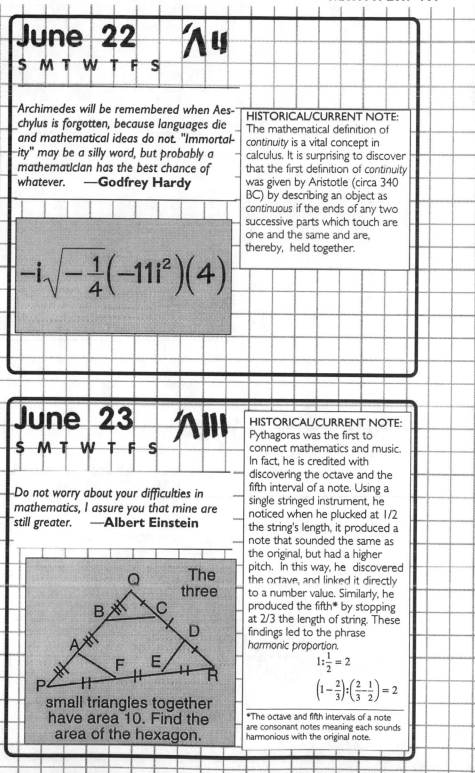

June 22

S M T W T F S

∧ᴂ

Archimedes will be remembered when Aeschylus is forgotten, because languages die and mathematical ideas do not. "Immortality" may be a silly word, but probably a mathematician has the best chance of whatever. —**Godfrey Hardy**

HISTORICAL/CURRENT NOTE:
The mathematical definition of *continuity* is a vital concept in calculus. It is surprising to discover that the first definition of *continuity* was given by Aristotle (circa 340 BC) by describing an object as *continuous* if the ends of any two successive parts which touch are one and the same and are, thereby, held together.

$$-i\sqrt{-\frac{1}{4}(-11i^2)(4)}$$

June 23

S M T W T F S

∧Ⅲ

Do not worry about your difficulties in mathematics, I assure you that mine are still greater. —**Albert Einstein**

The three small triangles together have area 10. Find the area of the hexagon.

HISTORICAL/CURRENT NOTE:
Pythagoras was the first to connect mathematics and music. In fact, he is credited with discovering the octave and the fifth interval of a note. Using a single stringed instrument, he noticed when he plucked at 1/2 the string's length, it produced a note that sounded the same as the original, but had a higher pitch. In this way, he discovered the octave, and linked it directly to a number value. Similarly, he produced the fifth* by stopping at 2/3 the length of string. These findings led to the phrase *harmonic proportion*.

$$1:\frac{1}{2} = 2$$

$$\left(1-\frac{2}{3}\right):\left(\frac{2}{3}-\frac{1}{2}\right) = 2$$

*The octave and fifth intervals of a note are consonant notes meaning each sounds harmonious with the original note.

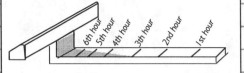

Rendition of the Egyptian sundial found about 1500 BC.

June 24

S M T W T F S

Λ ɰɰ

If others would but reflect on mathematical truths as deeply and as continuously as I have, they would make my discoveries.
—**Karl Gauss**

HISTORICAL/CURRENT NOTE:
Using their mathematics and astronomy, the Egyptians developed a means of keeping time utilizing the sun. Today, the oldest existing sundial is housed in the Berlin Museum, and it is an Egyptian time piece. They designed it so that the shadow from 6am to 12pm shortens, and from 12pm to 6pm lengthens. This accounts for the division of a day into 12 hour intervals.

$$\log_{10}\left[\frac{100,000}{1\,(\log_{137}137)}\right]$$

June 25

S M T W T F S

I keep the subject constantly before me and wait till the first dawnings open little by little into full light. —**Isaac Newton**

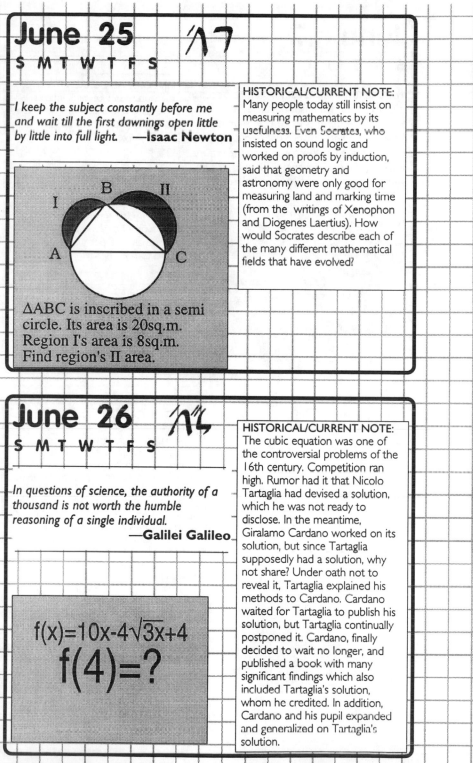

△ABC is inscribed in a semi circle. Its area is 20sq.m. Region I's area is 8sq.m. Find region's II area.

HISTORICAL/CURRENT NOTE: Many people today still insist on measuring mathematics by its usefulness. Even Socrates, who insisted on sound logic and worked on proofs by induction, said that geometry and astronomy were only good for measuring land and marking time (from the writings of Xenophon and Diogenes Laertius). How would Socrates describe each of the many different mathematical fields that have evolved?

June 26

S M T W T F S

In questions of science, the authority of a thousand is not worth the humble reasoning of a single individual.
—**Galilei Galileo**

$$f(x)=10x-4\sqrt{3x}+4$$
$$f(4)=?$$

HISTORICAL/CURRENT NOTE: The cubic equation was one of the controversial problems of the 16th century. Competition ran high. Rumor had it that Nicolo Tartaglia had devised a solution, which he was not ready to disclose. In the meantime, Giralamo Cardano worked on its solution, but since Tartaglia supposedly had a solution, why not share? Under oath not to reveal it, Tartaglia explained his methods to Cardano. Cardano waited for Tartaglia to publish his solution, but Tartaglia continually postponed it. Cardano, finally decided to wait no longer, and published a book with many significant findings which also included Tartaglia's solution, whom he credited. In addition, Cardano and his pupil expanded and generalized on Tartaglia's solution.

June 27
S M T W F S

The scientist does not study nature because it is useful; he studies it because he delights in it, and he delights in it because it is beautiful. If nature were not beautiful, it would not be worth knowing, and if nature were not worth knowing, life would not be worth living. —**Henry Poincaré**

HISTORICAL/CURRENT NOTE: Measurements of bank interest rates and tax rates were the main uses for percents in the 16th century, even as they are today. In the 17th century, the rate was usually expressed as hundredths. Percent signs began to emerge in the 15th century as *per c̊* or *p c̊,* which were derived from *per cento.* By the mid 17th century the symbol evolved to *per ÷,* and eventually the per was dropped. And today it is **%**.

$$\frac{(631_?)}{100_{two}} = 79_{ten}$$

June 28
S M T W F S

When we try to pick out anything by itself we find it hitched to everything else in the universe. —**John Muir**

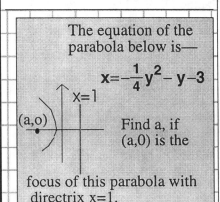

The equation of the parabola below is—

$$x = -\frac{1}{4}y^2 - y - 3$$

(a,0) X=1

Find a, if (a,0) is the focus of this parabola with directrix x=1.

HISTORICAL/CURRENT NOTE: There are hundreds, if not thousands, of mathematical recreation books, whose problems are based on a rich and old legacy. Some problems date back to the ancient Egyptians, Greeks and Asians. They were designed both for practical purposes and to tantalize the intellect. The treasure trove of puzzle problems has grown over the centuries. Many of these have been recycled and embellished by today's conundrums writers. Some had to do with finding the time on a sundial, arithmetic series and Diophantine equations. Among the sets of problems written during Medieval times are those by Alcuin of York (circa 775) which were sent to Charlemagne, by Rabbi ben Erza (circa 1140), by Fibonacci (1202) and by Bhāskara. Other mathematical recreation books from the past include Claude-Gasper Bachet's *Problems plaisans & delectables* of 1612; the Japanese work *Sampō Benran*) by Takeda Shingens of 1824.

June 29

S M T W T F S

As long as a branch of science offers an abundance of problems, so long is it alive.
—**David Hilbert**

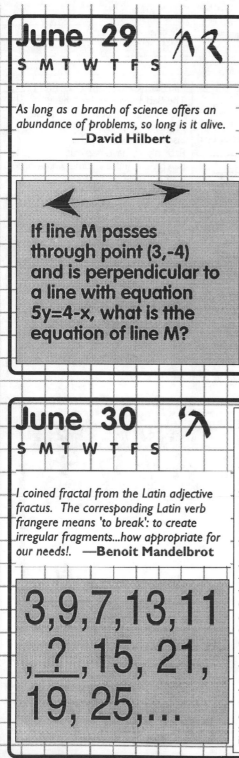

If line M passes through point (3,-4) and is perpendicular to a line with equation 5y=4-x, what is tthe equation of line M?

HISTORICAL/CURRENT NOTE:
While the Roman empire grew in military might, riches, and territories, the progress of mathematics suffered. The Romans were more concerned with the practical uses of mathematics than its theory and the discovery of new ideas. Consequently, their innovative mathematical ideas appear in such things as aqueducts, the Roman arch and surveying techniques. Although the Romans created a goddess named Numeraria, she was devoted to wealth and its acquisition rather than new ideas involving numbers and mathematics. Cicero said geometry held a high place of honor among the Greeks, but unfortunately was not appreciated by the Romans. All this may explain why many Latin scholars studied outside of Italy.

June 30

S M T W T F S

I coined fractal from the Latin adjective fractus. The corresponding Latin verb frangere means 'to break': to create irregular fragments...how appropriate for our needs!. —**Benoit Mandelbrot**

3,9,7,13,11
,_?_, 15, 21,
19, 25,...

HISTORICAL/CURRENT NOTE:
With the ever growing connectivity of computers via the Internet comes the perils of viruses. Today, a computer virus can spread around the world in a matter of hours. Rather than looking to mathematics and science for these problems, computer scientists are exploring the functioning of the human immune system to devise more effective computer solutions, such as computer antibodies or antiviral programs, to design adigital immune systems for the computer. One approach is led by the IBM Thomas J. Watson Research Center in New York. Another method is being explored at the University of New Mexico in Albuquerque by developing a means for the computer to know "itself" and how "it" should work. Thus, by being able to recognize changes inconsistent with its notion of "self", it may be able to identify an intruder virus behavior. In both cases the human immune system is serving as a model for the computer immune system.

July 1

S M T W T F S

July dates are written in the Chinese script number system

It isn't that they can't see the solution. It is that they can't see the problem.
—G.K. Chesterton

HISTORICAL/CURRENT NOTE:
The Babylonians did not have readily available access to marble or stone to make permanent inscriptions. Nor did the early Babylonians have papyrus or parchment. Initially some wrote on leather, while those in the northern area used clay and a cuneiform stylus (wedged shaped stick) to write information, problems and computations. These were then baked in a kiln or dried by the sun and became their permanent records. It was these clay tablets which furnished information about their number system, computation, and everyday records.

Here is a sequence of patterns. Determine what goes in the blank box.

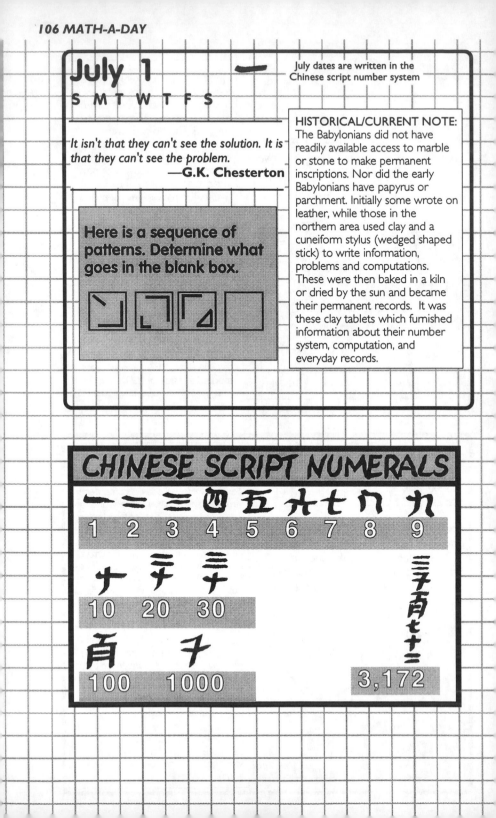

CHINESE SCRIPT NUMERALS

一	二	三	四	五	六	七	八	九
1	2	3	4	5	6	7	8	9

十	干	卅
10	20	30

百	千
100	1000

3,172

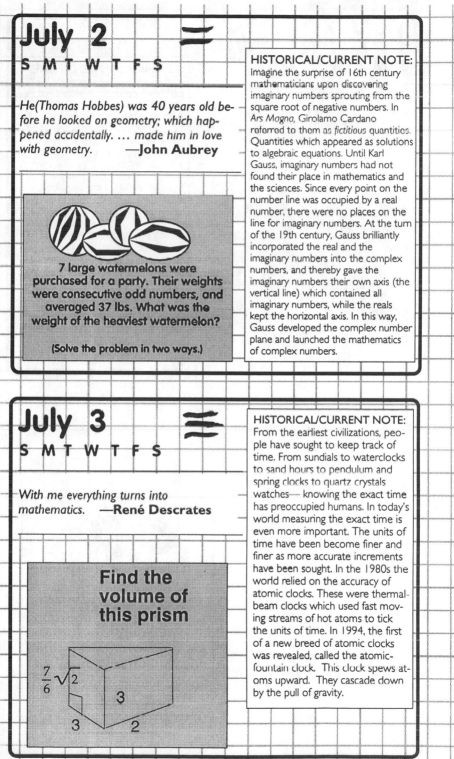

July 2

S M T W T F S

He(Thomas Hobbes) was 40 years old be-fore he looked on geometry; which hap-pened accidentally. ... made him in love with geometry. —**John Aubrey**

7 large watermelons were purchased for a party. Their weights were consecutive odd numbers, and averaged 37 lbs. What was the weight of the heaviest watermelon?

(Solve the problem in two ways.)

HISTORICAL/CURRENT NOTE:
Imagine the surprise of 16th century mathematicians upon discovering imaginary numbers sprouting from the square root of negative numbers. In *Ars Magna*, Girolamo Cardano referred to them as *fictitious* quantities. Quantities which appeared as solutions to algebraic equations. Until Karl Gauss, imaginary numbers had not found their place in mathematics and the sciences. Since every point on the number line was occupied by a real number, there were no places on the line for imaginary numbers. At the turn of the 19th century, Gauss brilliantly incorporated the real and the imaginary numbers into the complex numbers, and thereby gave the imaginary numbers their own axis (the vertical line) which contained all imaginary numbers, while the reals kept the horizontal axis. In this way, Gauss developed the complex number plane and launched the mathematics of complex numbers.

July 3

S M T W T F S

With me everything turns into mathematics. —**René Descrates**

Find the volume of this prism

$\frac{7}{6}\sqrt{2}$

3

3   2

HISTORICAL/CURRENT NOTE:
From the earliest civilizations, peo-ple have sought to keep track of time. From sundials to waterclocks to sand hours to pendulum and spring clocks to quartz crystals watches— knowing the exact time has preoccupied humans. In today's world measuring the exact time is even more important. The units of time have been become finer and finer as more accurate increments have been sought. In the 1980s the world relied on the accuracy of atomic clocks. These were thermal-beam clocks which used fast mov-ing streams of hot atoms to tick the units of time. In 1994, the first of a new breed of atomic clocks was revealed, called the atomic-fountain clock. This clock spews at-oms upward. They cascade down by the pull of gravity.

July 4
S M T W T F S

The human mind has first to construct forms, independently, before we can find them in things. —**Albert Einstein**

MMDXIV minus

Roman

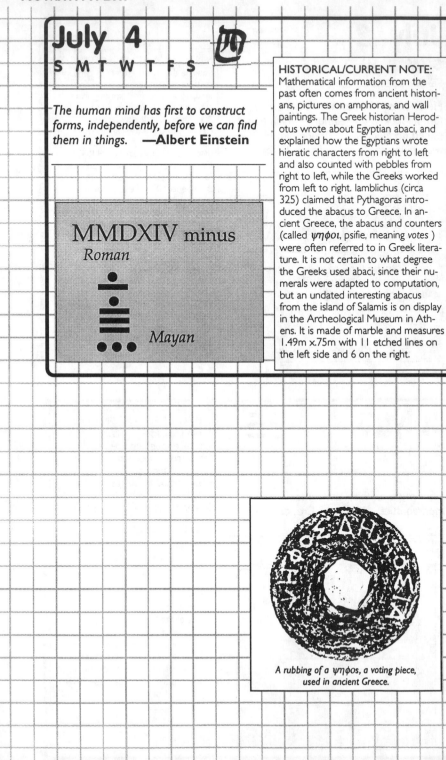

Mayan

HISTORICAL/CURRENT NOTE:
Mathematical information from the past often comes from ancient historians, pictures on amphoras, and wall paintings. The Greek historian Herodotus wrote about Egyptian abaci, and explained how the Egyptians wrote hieratic characters from right to left and also counted with pebbles from right to left, while the Greeks worked from left to right. Iamblichus (circa 325) claimed that Pythagoras introduced the abacus to Greece. In ancient Greece, the abacus and counters (called ψηφοι, psifie, meaning *votes*) were often referred to in Greek literature. It is not certain to what degree the Greeks used abaci, since their numerals were adapted to computation, but an undated interesting abacus from the island of Salamis is on display in the Archeological Museum in Athens. It is made of marble and measures 1.49m x.75m with 11 etched lines on the left side and 6 on the right.

A rubbing of a ψηφos, a voting piece, used in ancient Greece.

July 5

S M T W T F S

五

Unfortunately what is little recognized is that the most worthwhile scientific books are those in which the author clearly indicates what he does not know; for an author most hurts his readers by concealing difficulties. —**Evariste Galois**

$$\frac{6.25}{\sqrt{.0625}}$$

HISTORICAL/CURRENT NOTE:
One of the most important tools of mathematics is logic. Logic is timeless and universal, as witness this quote, over 2500 years old, from the *Anuradha Sutra*: "Can the Tathagata be recognized through form?" the Buddah asked Anuradha. "No, master," Anuradha replied. "Can the Tathagata be recognized outside of form?" "No, master." "Can the Tathagata be recognized through feeling, perception, mental formations or consciousness?" "No, master." "Anuradha, you cannot find the Tathagata even in this life, why do you want to solve the problem of whether I will continue to exist or cease to exist, or both continue and cease to exist, or neither continue nor cease to exist after death?" This is analogous to what Robert Oppenheimer said in 1954—"If we ask, for instance, whether the position of an electron remains the same, we must say 'no;' if we ask whether the electron's position changes with time, we must say 'no;' if we ask whether the electron is at rest, we must say 'no;' if we ask whether it is in motion, we must say 'no.'

July 6

S M T W T F S

六

Anyone who cannot cope with mathematics is not fully human. At best he is a tolerable subhuman who has learned to wear shoes, bathe, and not make messes in the house. —**Robert Heinlein**

$$\left(\frac{3+3}{3^3}\right)\left(\frac{3\left(3+3^1-3^0\right)3^2}{3-3^0}\right)$$

HISTORICAL/CURRENT NOTE:
An old problem dealing with inheritance dates back centuries and across cultures. *A man plans to bequeath his estate to his widow and to the child his wife is expecting. If it's a boy the son gets 2/3 and his wife 1/3 of the estate. If it's a girl, the daughter gets 1/3 and the wife 2/3. The wife has twins, a boy and a girl. How is the estate divided?* Similar problems appear in the works of the Middle Ages, changing the offspring to triplets (2 boys and a girl), which gives a new twist to the problem.

July 7
S M T W F S

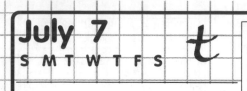

Mathematics knows no races or geographic boundaries; for mathematics, the cultural world is one country.—**David Hilbert**

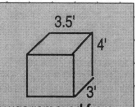

A weight was removed from this vat of water. The water level sank 4". If water weighs about 8 lbs per cubic foot, how many pounds is the weight?

HISTORICAL/CURRENT NOTE:
How did the Web originate?
Foundations for the Web were laid in the 1940s with the exploration of electronic means to store, organize and disseminate information. Many intermediate stages had to occur before the Web took off—these include work by Vannevar Bush, Douglas Engelbart, Charles Goldfarb, Alan Kay, Ted Nelson, and especially Tim Bernes-Lee. Bernes-Lee is responsible for creating the World Wide Web, which he did while working for CERN (Conseil Européen pour la Recherche Nucléaire), the European Laborabory for Particle Physics. In 1987 U.S. and CERN laboratories connected to the Internet, thus providing a rapid means to share and exchange ideas. It was in 1989 that Bernes-Lee proposed a hypertext system which was refined by Bernes-Lee and Robert Cailiau in 1990. Bernes-Lee's proposal included programming, which enabled any user to access information from any type of computer.

July 8
S M T W F S

$$\int_{1}^{\infty} \frac{1}{x}\,dx$$

To understand this for sense it is not required that a man should be a geometrician or a logician, but that he should be mad.
—**Thomas Hobbes**

$$|AB|=?$$

A
3
3   4
4   5
5   4
B

HISTORICAL/CURRENT NOTE:
The fame and influence of Pythagoras spans centuries. Pythagoras even appears in Shakespeare's *Merchant of Venice*, when reference is made to the society's belief in the transmigration of souls:
"Thou almost mak'st me waver in my faith, to hold opinion with Pythagoras, that souls of animals infuse themselves into the trunks of men."

July 9

S M T W T F S

Geometry enlightens the intellect and sets one's mind right. —**Ibn Khaldun**

If |AP|=17
and |PB|=6,
then |AD|=?

HISTORICAL/CURRENT NOTE: Herodotus (c.484-425 BC) describes how Rameses II divided the land of Egypt into equally sized square plots for all inhabitants, and how revenues from these plots were collected in the form of rents. If the flooding Nile removed soil from a person's land, the damage was assessed and the rent adjusted accordingly. Herodotus claims that this is how geometry was introduced in Egypt. It is also interesting to note that Egyptian priests plotted the path of the star Sirius in order to predict the annual flooding of the Nile.

July 10

S M T W T F S

"You know, you can think of almost everything as a math problem."
—**Jon Scieszka**
Math Curse

The trickster

Tom secretly put two black marbles in a bag, and told Ann it contained one black and one white. Tom boldly insisted that Ann would not pick the white marble on her first draw. What did Ann say to Tom in order to out trick him?

HISTORICAL/CURRENT NOTE: Today supercomputers and sophisticated methods (e.g. quadratic sieve and a number field sieve) are used to search for larger and larger prime numbers and prime factors of numbers. It is startling to realize that about 230 BC someone was interested in prime numbers. Eratosthenes, the ancient Greek who determined the circumference of the equator to within 2% of its value, was also interested in devising a system to sift out prime numbers. He devised a number sieve in which from a list of, for example, the first 100 natural numbers, he crossed out all those that were multiples of 2, 3, 5, 7,... and the numbers left were prime.

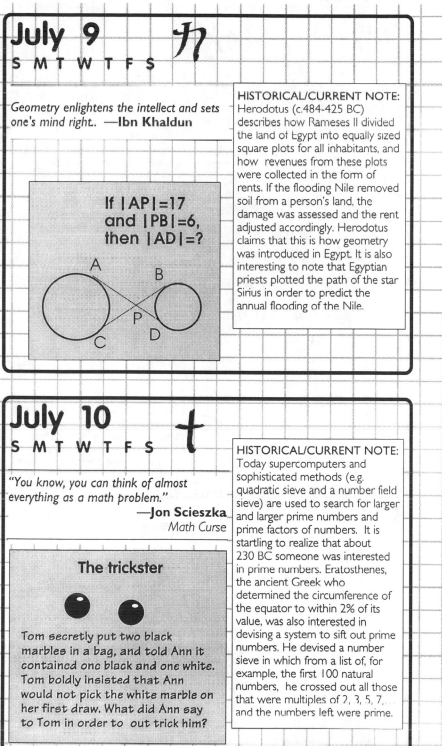

July 11
S M T W T F S

Number theorists are like lotus-eaters – having once tasted of this food they can never give it up.
—Leopold Kronecker

If points (3,-5), (0,-7), and (16.5, a) are collinear, then determine the value of a.

HISTORICAL/CURRENT NOTE: The expression "thrown for a loop" may have a mathematical connection. The "infinite loop" is the nemesis of a computer program. In the process of writing a program, an error or oversight may take place which at a specific place or set of commands triggers a set of instructions that makes a procedure continually repeat itself in an endless circle. The only way to stop it is to turn the program off.

July 12
S M T W T F S

He uses statistics as a drunken man uses lamp posts – for support rather than illumination. **—Andrew Lang**

The measure of arc AMB=60°. Its length is (2 2/3)π. Find |OB|.

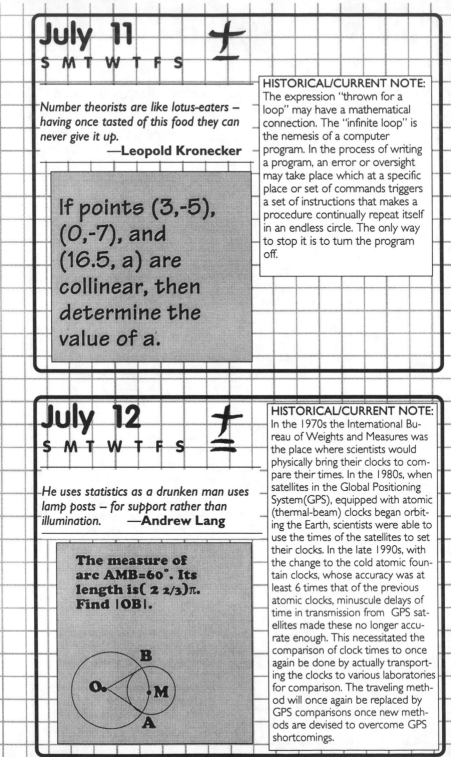

HISTORICAL/CURRENT NOTE: In the 1970s the International Bureau of Weights and Measures was the place where scientists would physically bring their clocks to compare their times. In the 1980s, when satellites in the Global Positioning System(GPS), equipped with atomic (thermal-beam) clocks began orbiting the Earth, scientists were able to use the times of the satellites to set their clocks. In the late 1990s, with the change to the cold atomic fountain clocks, whose accuracy was at least 6 times that of the previous atomic clocks, minuscule delays of time in transmission from GPS satellites made these no longer accurate enough. This necessitated the comparison of clock times to once again be done by actually transporting the clocks to various laboratories for comparison. The traveling method will once again be replaced by GPS comparisons once new methods are devised to overcome GPS shortcomings.

July 13
S M T W T F S

...She knew only that if she did or said *thus-and-so*, men would unerringly respond with the complimentary *thus-and-so*. It was like a mathematical formula and no more difficult, for mathematics was the one subject that had come easy to Scarlett in her school days. **—Margaret Mitchell**
Gone With The Wind

$$\frac{\text{3 septillion}}{10^{23}}$$

HISTORICAL/CURRENT NOTE: How time flies! The Julian calendar was off a mere 11 minutes 14 seconds annually, yet by the year 1572, this discrepancy had accumulated to 10 days. It took a papal edict by Gregory XIII and the work of two mathematician-astronomers to come up with a solution — the Gregorian calendar still in use today. On February 24, 1582, the new calendar was instated. It deleted the 10 extra days by allowing the days between October 4, 1582 and October 15th to be skipped. From then on leap years were every four years, and only century years divisible by 400 were allowed to be a leap year. For religious differences, the Gregorian calendar was not immediately adopted by non-Catholic countries. In fact, Britain did not adopt it until 1752, Sweden in 1753, Protestant German states in 1700, and Eastern Europe and the USSR held out until the 20th century.

July 14
S M T W T F S

...there is a God precisely because Nature itself, even in chaos, cannot proceed except in an orderly and regular manner.
—Immanuel Kant

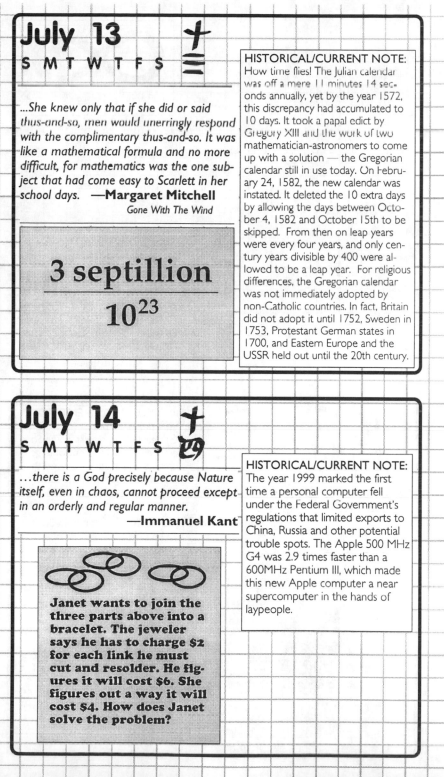

Janet wants to join the three parts above into a bracelet. The jeweler says he has to charge $2 for each link he must cut and resolder. He figures it will cost $6. She figures out a way it will cost $4. How does Janet solve the problem?

HISTORICAL/CURRENT NOTE: The year 1999 marked the first time a personal computer fell under the Federal Government's regulations that limited exports to China, Russia and other potential trouble spots. The Apple 500 MHz G4 was 2.9 times faster than a 600MHz Pentium III, which made this new Apple computer a near supercomputer in the hands of laypeople.

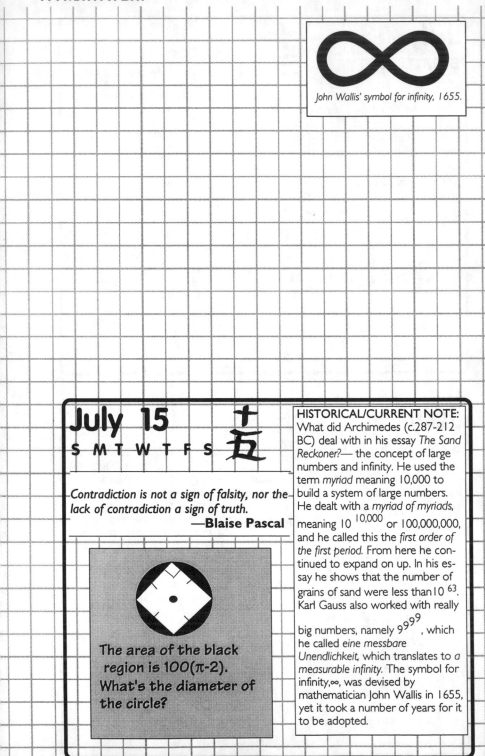

John Wallis' symbol for infinity, 1655.

July 15

S M T W T F S

十
五

Contradiction is not a sign of falsity, nor the lack of contradiction a sign of truth.

—Blaise Pascal

The area of the black region is $100(\pi-2)$. What's the diameter of the circle?

HISTORICAL/CURRENT NOTE: What did Archimedes (c.287-212 BC) deal with in his essay *The Sand Reckoner?*— the concept of large numbers and infinity. He used the term *myriad* meaning 10,000 to build a system of large numbers. He dealt with a *myriad of myriads,* meaning $10^{10,000}$ or 100,000,000, and he called this the *first order of the first period.* From here he continued to expand on up. In his essay he shows that the number of grains of sand were less than 10^{63}. Karl Gauss also worked with really big numbers, namely $9^{9^{9^9}}$, which he called *eine messbare Unendlichkeit,* which translates to *a measurable infinity.* The symbol for infinity, ∞, was devised by mathematician John Wallis in 1655, yet it took a number of years for it to be adopted.

July 16

S M T W T F S

Mathematics, as much as music or any other art, is one of the means by which we rise to a complete self-consciousness. The significance of mathematics resides precisely in the fact that it is an art; by informing us of the nature of our own minds it informs us of much that depends on our minds. —**John W. N. Sullivan**

YOUR CHANCES

If (1/100)% of the population can get a rare type of flu, then how many people in 100,000 can catch this flu?

HISTORICAL/CURRENT NOTE:
In 1838 artist and inventor Samuel Morse devised the Morse Code, which was modified for the International Morse Code in 1851. Dots and dash combinations are used to write words and numbers. Here is how the digits from 0 through 9 are written—

0 is – – – – –
1 is • – – – –
2 is • • – – –
3 is • • • – –
4 is • • • • –
5 is • • • • •
6 is – • • • •
7 is – – • • •
8 is – – – • •
9 is – – – – •
period is written • – • – • –
comma is – – • • – –
So 459 would be written as
• • • • –   • • • • •   – – – – •

July 17

S M T W T F S

He who can properly define and divide is to be considered a god. —**Plato**

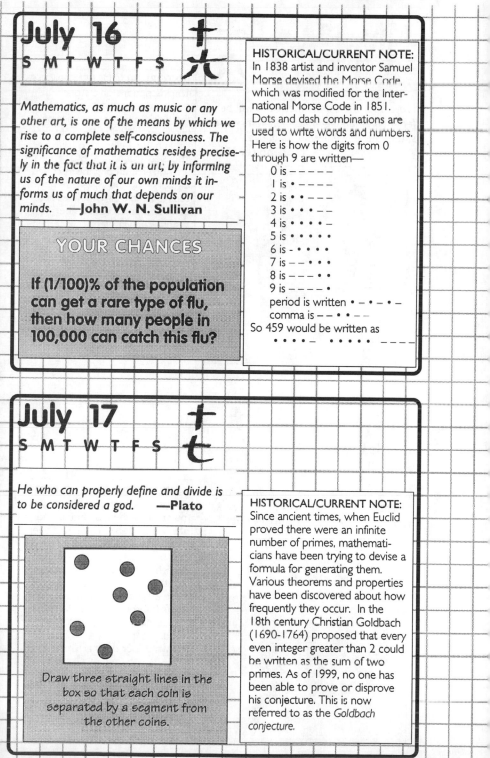

Draw three straight lines in the box so that each coin is separated by a segment from the other coins.

HISTORICAL/CURRENT NOTE:
Since ancient times, when Euclid proved there were an infinite number of primes, mathematicians have been trying to devise a formula for generating them. Various theorems and properties have been discovered about how frequently they occur. In the 18th century Christian Goldbach (1690-1764) proposed that every even integer greater than 2 could be written as the sum of two primes. As of 1999, no one has been able to prove or disprove his conjecture. This is now referred to as the *Goldbach conjecture*.

July 18

S M T W T F S

Prayers for the condemned man will be offered on an adding machine. Numbers constitute the only universal language.
—**Nathanael West**

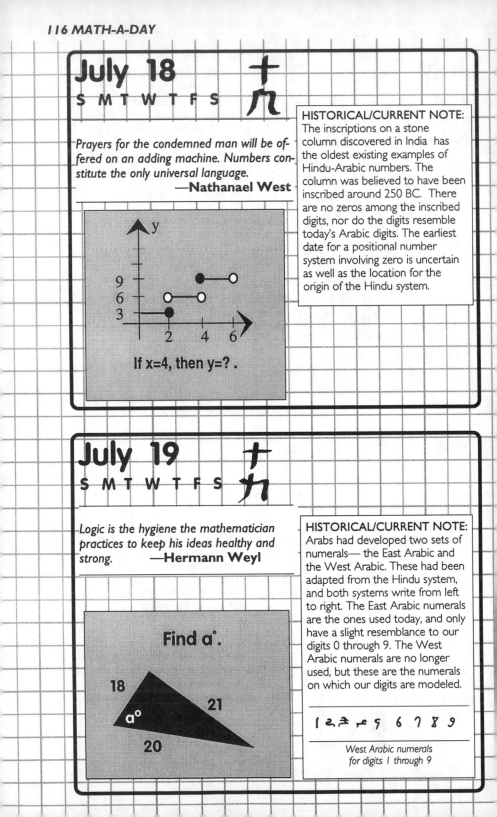

If x=4, then y=? .

HISTORICAL/CURRENT NOTE:
The inscriptions on a stone column discovered in India has the oldest existing examples of Hindu-Arabic numbers. The column was believed to have been inscribed around 250 BC. There are no zeros among the inscribed digits, nor do the digits resemble today's Arabic digits. The earliest date for a positional number system involving zero is uncertain as well as the location for the origin of the Hindu system.

July 19

S M T W T F S

Logic is the hygiene the mathematician practices to keep his ideas healthy and strong.   —**Hermann Weyl**

Find a°.

18
21
a°
20

HISTORICAL/CURRENT NOTE:
Arabs had developed two sets of numerals— the East Arabic and the West Arabic. These had been adapted from the Hindu system, and both systems write from left to right. The East Arabic numerals are the ones used today, and only have a slight resemblance to our digits 0 through 9. The West Arabic numerals are no longer used, but these are the numerals on which our digits are modeled.

1 2 ⋧ ⋉ 9 6 7 8 9

*West Arabic numerals
for digits I through 9*

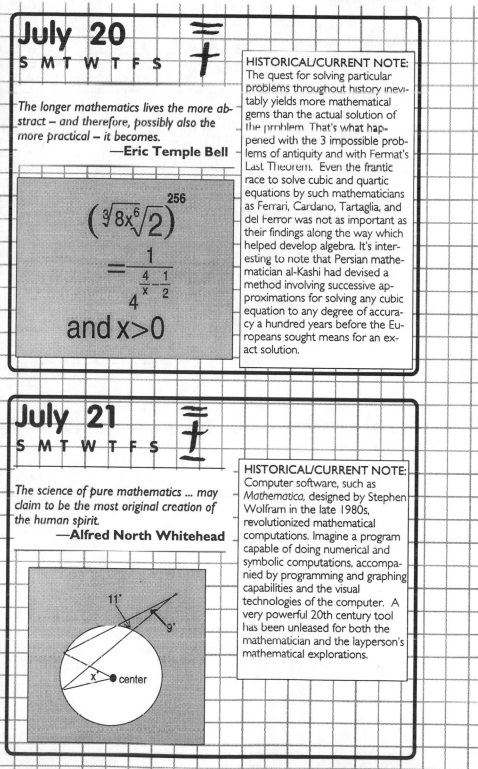

July 20

S M T W T F S

The longer mathematics lives the more abstract – and therefore, possibly also the more practical – it becomes.
—**Eric Temple Bell**

$$\left(\sqrt[3]{8x^6}\sqrt{2}\right)^{256}$$

$$= \frac{1}{4^{\frac{4}{x}-\frac{1}{2}}}$$

and x>0

HISTORICAL/CURRENT NOTE:
The quest for solving particular problems throughout history inevitably yields more mathematical gems than the actual solution of the problem. That's what happened with the 3 impossible problems of antiquity and with Fermat's Last Theorem. Even the frantic race to solve cubic and quartic equations by such mathematicians as Ferrari, Cardano, Tartaglia, and del Ferror was not as important as their findings along the way which helped develop algebra. It's interesting to note that Persian mathematician al-Kashi had devised a method involving successive approximations for solving any cubic equation to any degree of accuracy a hundred years before the Europeans sought means for an exact solution.

July 21

S M T W T F S

The science of pure mathematics ... may claim to be the most original creation of the human spirit.
—**Alfred North Whitehead**

11°

9°

x° center

HISTORICAL/CURRENT NOTE:
Computer software, such as *Mathematica*, designed by Stephen Wolfram in the late 1980s, revolutionized mathematical computations. Imagine a program capable of doing numerical and symbolic computations, accompanied by programming and graphing capabilities and the visual technologies of the computer. A very powerful 20th century tool has been unleased for both the mathematician and the layperson's mathematical explorations.

July 22
S M T W T F S

The time it took the sphere to swing from end to end was determined by an arcane conspiracy between the most timeless of measures: the singularity of the point of suspension, the duality of the plane's dimensions, the triadic beginning of π, the secret quadratic nature of the root, and the unnumbered perfection of the circle itself. —**Umberto Eco**
Foucault's Pendulum

$$\frac{\dfrac{4x-\dfrac{1}{x}}{1-\dfrac{2}{2x+1}}}{\dfrac{4x^2+4x+1}{3x}}$$

HISTORICAL/CURRENT NOTE:
In 1993 Andrew Wiles, presented his first proof of *Fermat's Last Theorem (FLT)* [that for any natural number, $n>2$, $x^n+y^n=z^n$ has no positive integral solutions]. With the help of Richard Taylor a revision was submitted in 1995, and has been given the nod. Some historic steps in the quest of FLT include— in 1753 Euler proved FLT for $n=4$ and Gauss corrected his attempt for $n=3$, in 1825 Legendre proved FLT for $n=5$, in 1839 Lamé proved it for $n=7$, in the 18th century Germain added additional values of n to the proven list, and D. Lehmer and E. Lehmer expanded her work. In the 20th century, Joe Butler and Richard Crandall brought computer technology to FLT and proved it for all exponents $n<4{,}000{,}000$.

July 23
S M T W T F S

Of all the great men of antiquity, Archimedes may be the one who most deserves to be placed beside Homer.
—**Jean le Rond D'Alembert**

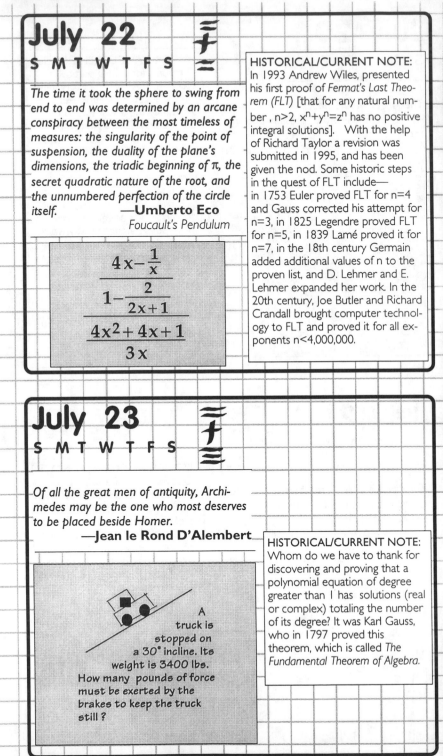

A truck is stopped on a 30° incline. Its weight is 3400 lbs. How many pounds of force must be exerted by the brakes to keep the truck still?

HISTORICAL/CURRENT NOTE:
Whom do we have to thank for discovering and proving that a polynomial equation of degree greater than 1 has solutions (real or complex) totaling the number of its degree? It was Karl Gauss, who in 1797 proved this theorem, which is called *The Fundamental Theorem of Algebra*.

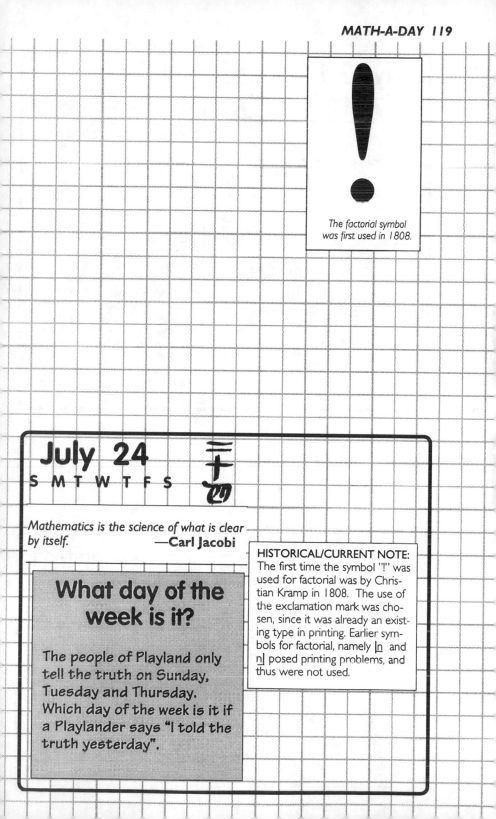

*The factorial symbol
was first used in 1808.*

July 24
S M T W T F S

*Mathematics is the science of what is clear
by itself.* —**Carl Jacobi**

What day of the week is it?

The people of Playland only
tell the truth on Sunday,
Tuesday and Thursday.
Which day of the week is it if
a Playlander says "I told the
truth yesterday".

HISTORICAL/CURRENT NOTE:
The first time the symbol "!" was
used for factorial was by Christian Kramp in 1808. The use of
the exclamation mark was chosen, since it was already an existing type in printing. Earlier symbols for factorial, namely ⌊n and
n⌋ posed printing problems, and
thus were not used.

July 25
S M T W T F S 三五

Whoever despises the high wisdom of mathematics nourishes himself on delusion. —**Leonardo da Vinci**

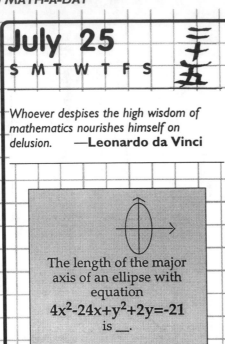

The length of the major axis of an ellipse with equation
$$4x^2-24x+y^2+2y=-21$$
is ___.

HISTORICAL/CURRENT NOTE:
Mathematical modeling has taken to the roads. The science of *traffic-flow* is being explored by mathematicians and scientists all over the world. Researchers are now looking at traffic as a non-linear dynamic system, that is, as a complex system. Some have designed virtual models around cellular automata. The computer monitor becomes the terrain for roads, the pixels the vehicles; mix in a set of simple rules, and voilá traffic is flowing along virtual roads. Some scientists are comparing traffic to the movement of molecules, and others to the flow of sand, while traditional traffic researchers are questioning such methods. Only time will tell which method or methods will pan out in this ever increasing problem.

July 26
S M T W T F S 二十六

Mathematicians have tried in vain to this day to discover some order in the sequence of prime numbers, and we have reason to believe that it is a mystery into which the human mind will never penetrate.. — **Leonhard Euler**

Speed Limit

On a freeway, two cars are going the same direction. The car traveling 65mph passes the car going 55mph. Assuming constant speeds, after 108 miles, how many minutes ahead is the 65mph car?

HISTORICAL/CURRENT NOTE:
Where did the idea of symbolic logic originate? Since arithmetic errors are usually easy to detect, Gottfried Leibniz (1646-1716) decided to use this idea to try to devise a language which could detect logic errors which would appear as arithmetic errors.
In his publication *Dissertatio de arte combinatoria,* Leibniz set up an arithmetic system to make logical deductions. By so doing, he set the ground work for symbolic logic.

July 27
S M T W T F S

Mathematics is the cheapest science. Unlike physics or chemistry, it does not require any expensive equipment. All one needs for mathematics is a pencil and paper. —**George Polya**

The curious cube

The first odd perfect cube greater than 1 is _?_ , and when cubed the sum of its digits is _?_ .

HISTORICAL/CURRENT NOTE: Today, any students taking engineering, chemistry, physics, or advanced mathematics courses carry a sophisticated, yet compact calculator, jam packed with all the necessary tables, functions and operations and probably graphing capabilities. Yet, in the 1960s the slide rule was the tool that such students had available. The slide rule's origins date back to the 17th century. In 1620, British mathematician Edmund Gunter, who designed and constructed navigation instruments, made a ruler marked with logarithmic distances. Using compasses with the ruler, numbers could be multiplied, and divided by simply adding and subtracting numbers. Just one year later, mathematician William Oughtred invented the rectilinear slide rule with two sliding log scales. He also devised a circular slide rule with two circular sliding scales.

July 28
S M T W T F S

Mathematics is written for mathematicians. —**Nicholas Copernicus**

4.96

10

Without cutting this sheet of paper, simply fold it to form lateral faces of a square based prism. Assuming top and bottom are added later from other paper, what is the maximum volume of this prism?

HISTORICAL/CURRENT NOTE: Mathematics is becoming more pervasive everyday. Witness the fragrance by Givenchy called π (pi) and marketed as *a sign of intelligent life*. Isn't it interesting that it was developed only for men?

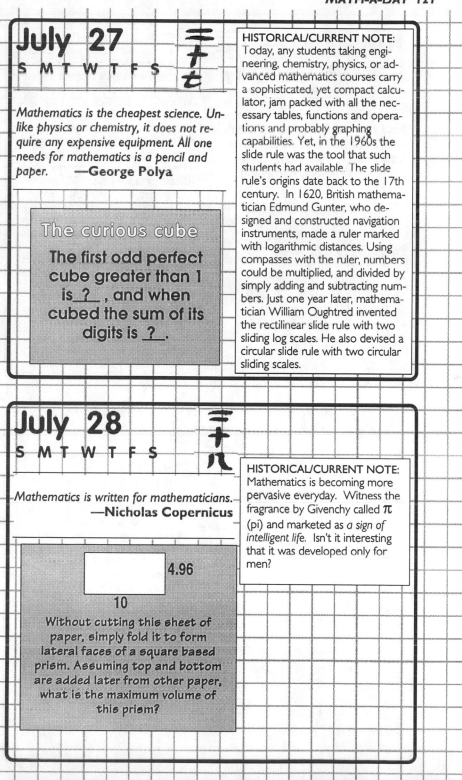

July 29

S M T W T F S

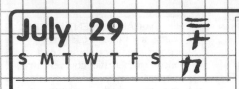

I had a feeling once about Mathematics - that I saw it all. Depth beyond depth was revealed to me - the Byss and Abyss. ... but it was after dinner and I let it go.
—**Winston Churchill**

A sphere and a right circular cylinder have the same height, $22\sqrt{6}$, and the same volume. What is the radius of the cylinder?

HISTORICAL/CURRENT NOTE:
The story of the slide rule did not end in the 1600s, but continued through the 1940s. Richard Delamain, a pupil of Oughtred, published a booklet on circular slide rules, of which he claimed to be the innovator. In 1632, Oughtred published a booklet about his circular slide rule. The two men began a feud over its priority which lasted until the death of Delamain. Oughtred is definitely the creator of the rectilinear slide rule, which resembles today's C and D scale slide rules. In 1815, Peter Mark Roget (author of Roget's Thesaurus) added the loglog scales. Over the years the accuracy of the slide rule improved with better materials and more accurate engraving instruments. In the 1840s new scales were added and both sides of the slide rule were utilized. It has now been replaced by the sophisticated hand held calculator.

July 30

S M T W T F S

How happy the lot of the mathematician. He is judged solely by his peers, and the standard is so high that no colleague or rival can ever win a reputation he does not deserve. —**W.H. Auden**

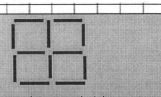

Relocate three toothpicks to new locations and end up with three squares the same size. Be sure to use each toothpick as a side of a square.

HISTORICAL/CURRENT NOTE:
The story behind the origin of *Venn diagrams* begins before John Venn (1834-1923) was born. It goes back to the 1600s when Leibniz used diagrams in symbolic logic. Sometimes *Venn diagrams* are referred to as *Euler diagrams*, because Euler is credited with creating them. But it was John Venn who improved the notations and encouraged their use to the point that they became a popular way in which to visualize sets and operations with sets.

July 31

S M T W T F S

"...I can see looming ahead one of those terrible exercises in probability where six men have white hats and six men have black hats and you have to work it out by mathematics how likely it is that the hats will get mixed up and in what proportion. If you start thinking about things like that, you would go round the bend. ..."

—Agatha Christie

An isosceles triangular vat has 83 1/3 cubic feet of water in it. What is the water's height in the tank?

4'

6'

10'

HISTORICAL/CURRENT NOTE:
It is amazing how the mind creates illusions to appear as reality. For example, it camouflages a blind spot, which all humans have, by filling it in with surrounding views. In 1934, artist Oscar Reutersvärd created an *impossible triangle*. Reutersvärd's work uses parallel lines which do not converge. This technique is often referred to as Japanese perspective. Over the years, his work and interest prompted other artists, mathematicians, and scientists to create and study impossible figures. Among these are Roger Penrose's *impossible tribar* and M.C. Escher's *Waterfall*. People never cease to be captivated by both optical illusions and impossible figures.

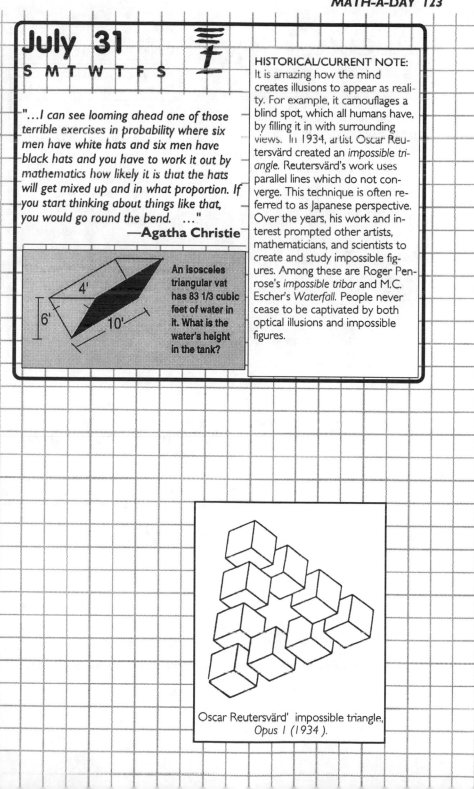

Oscar Reutersvärd' impossible triangle, *Opus 1 (1934)*.

August 1 א

S M T W T F S

August dates are written in the
Hebrew number system

*Mathematics is the art of giving the same
name to different things.*

—Henri Poincaré

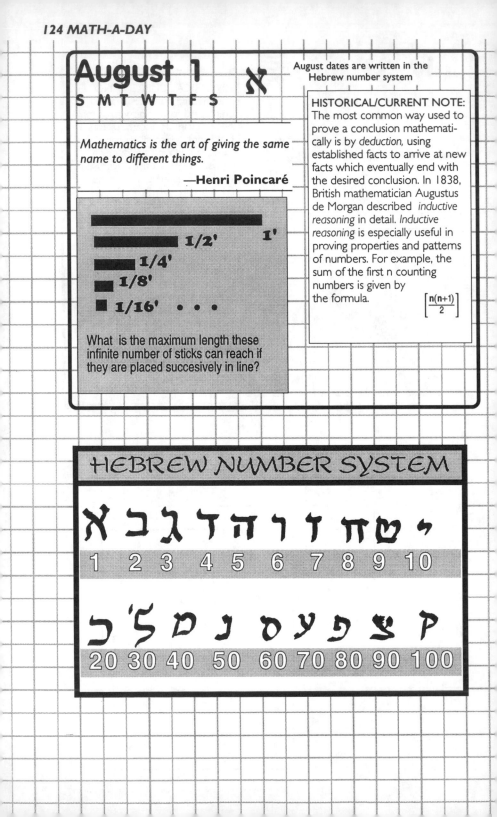

1'
1/2'
1/4'
1/8'
1/16' • • •

What is the maximum length these
infinite number of sticks can reach if
they are placed succesively in line?

HISTORICAL/CURRENT NOTE:
The most common way used to
prove a conclusion mathematically is by *deduction,* using
established facts to arrive at new
facts which eventually end with
the desired conclusion. In 1838,
British mathematician Augustus
de Morgan described *inductive
reasoning* in detail. *Inductive
reasoning* is especially useful in
proving properties and patterns
of numbers. For example, the
sum of the first n counting
numbers is given by
the formula. $\left[\frac{n(n+1)}{2}\right]$

HEBREW NUMBER SYSTEM

א	ב	ג	ד	ה	ו	ז	ח	ט	י
1	2	3	4	5	6	7	8	9	10

כ	ל	מ	נ	ס	ע	פ	צ	ק
20	30	40	50	60	70	80	90	100

August 2 ב
S M T W T F S

One cannot escape the feeling that these mathematical formulae have an independent existence and an intelligence of their own, that they are wiser than we are, wiser even that their discoverers, that we get more out of them than was originally put into them. — **Heinrich Hertz**

$$\left(\frac{13^2 - 13}{13} \right) + 13^0$$

HISTORICAL/CURRENT NOTE:
It is interesting to note that both mathematician David Hilbert (1862-1943) and philosopher Emmanuel Kant (1724-1804) were both from the city of Königsberg (now called Kaliningrad), the then city of the famous seven bridges problem. Euler's 1736 solution to the problem of walking over all seven bridges without doubling back was responsible for launching the field known today as topology. Using logic and networks(special graphs or drawings using arcs and vertices) Euler showed that the Königsberg bridge walk was impossible.

August 3 ג
S M T W T F S

Someone who began to read geometry with Euclid, when he had learned the first proposition, asked Euclid, "But what shall I get by learning these things?" Whereupon Euclid called his slave and said "Give him three pence since he must make gain out of what he learns." —**Stobaeus**

HISTORICAL/CURRENT NOTE:
The first known examples of indirect proofs date back to ancient Greece. Two famous examples are those of Euclid and the Pythagoreans. Euclid proved that *the number of primes is infinite* by assuming there were a finite number and showing such an assumption led to a contradiction. This was around 300 BC. Using the indirect method the Pythagoareans proved √2 was irrational around 500 BC. The indirect proof is also responsible, in an indirect way, for the discovery of non-Euclidean geometries.

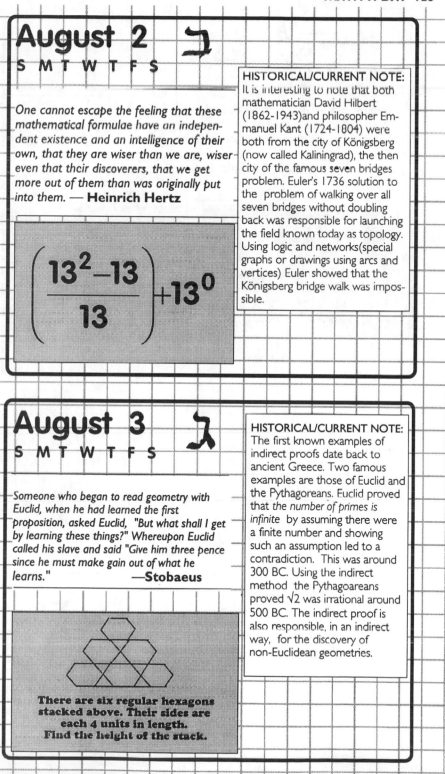

There are six regular hexagons stacked above. Their sides are each 4 units in length. Find the height of the stack.

August 4 ך
S M T W T F S

...creative mathematicians now, as in the past, are inspired by the art of mathematics rather than by any prospect of ultimate usefulness.
—Eric Temple Bell

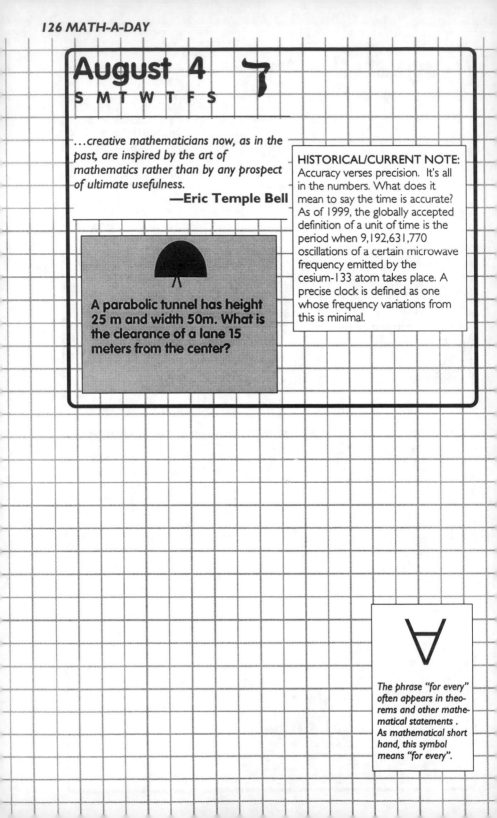

A parabolic tunnel has height 25 m and width 50m. What is the clearance of a lane 15 meters from the center?

HISTORICAL/CURRENT NOTE:
Accuracy verses precision. It's all in the numbers. What does it mean to say the time is accurate? As of 1999, the globally accepted definition of a unit of time is the period when 9,192,631,770 oscillations of a certain microwave frequency emitted by the cesium-133 atom takes place. A precise clock is defined as one whose frequency variations from this is minimal.

$\forall$

The phrase "for every" often appears in theorems and other mathematical statements . As mathematical short hand, this symbol means "for every".

August 5 ה
S M T W T F S

'Contrariwise,' continued Tweedledee, 'if it
was so, it might be; and if it were so, it
would be; but as it isn't, it ain't. That's
logic.' — **Lewis Carroll**
Alice in Wonderland

.0027+.000027
+.00000027+...
=what fraction

HISTORICAL/CURRENT NOTE:
With the advent of computers,
mathematics and mathematical
proofs entered a new era of problem solving. More problems are being tackled using computer force
than ever before. Programming and
problem analysis require keen logical diagnoses. Thus far, the
computer has been used mainly to
physically exhaust the possible cases
of a mathematical proof, as witness
the solution of the *Four-Color Map
Problem*. Will mathematics in the
new millenium rely too heavily on
this approach? Logic power and
computer power make a dynamic
duo, as long as logic is not sacrificed
for brute computer force.

August 6 ו
S M T W T F S

*This, therefore, is mathematics; she
reminds you of the invisible form of the
soul; she gives life to her own discoveries;
she awakens the mind and purifies the
intellect; she brings light to our intrinsic
ideas; she abolishes oblivion and ignorance
which are ours by birth..*
—**Proclus**

$$\left[\frac{5(x-y)}{y-x}\right]^2$$

HISTORICAL/CURRENT NOTE:
The *cardinals versus the ordinals*
sounds like a baseball billing. In
mathematics cardinal numbers denote the number of things in a set,
while the ordinal number indicates
the position of an object in a a set
or sequence. Thus, in a set of 5
objects, cardinal 5 denotes how
many objects, while ordinal 5th refers only to one particular object
of the set. Until Georg Cantor
(1845-1918) entered the world of
mathematics, there were no sets,
no cardinality of a set. Not only
did he introduce and develop set
theory, he is responsible for the
transfinite numbers— $\aleph_0$ aleph-
null, $\aleph_1$ aleph-one, $\aleph_2$,... which
are cardinal numbers for infinite
sets.

August 7

S M T W T F S

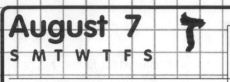

Of all things, good sense is the most fairly distributed: everyone thinks he is so well supplied with it that even those who are the hardest to satisfy in every other respect never desire more of it than they already have. —**René Descartes**

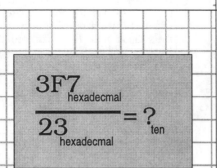

$$\frac{3F7_{\text{hexadecmal}}}{23_{\text{hexadecmal}}} = ?_{\text{ten}}$$

HISTORICAL/CURRENT NOTE:
What's the number **c** all about? It is a transfinite number that describes the cardinality of the set of real numbers (how many real numbers there are) between the number 0 and 1. **c** comes from the word *continuum* which refers to Cantor's *continuum hypothesis,* which states that any set of real numbers is either denumerable (can be counted by putting it into a one-to-one correspondence with the natural numbers) or can be put into a one-to-one correspondence with the real numbers between 0 and 1. Cantor was unable to prove or disprove his *continuum hypothesis.* In 1940 Kurt Gödel proved that Cantor's hypothesis *cannot be disproved.* In 1963, Paul Cohen showed the hypothesis to be *undecidable*— one of the three possibilities of a statement , i.e. true, false, or undecidable.

August 8

S M T W T F S

The moving power of mathematical invention is not reasoning but imagination. —**Augustus de Morgan**

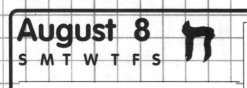

A rectangular sheet is formed into a right cylinder. If lids are placed on it, what would its volume?

HISTORICAL/CURRENT NOTE:
You'd think the Internet would have enough with its own traffic to contend with without adding road traffic to its Web. But that is just what many European countries are beginning to offer—visual traffic reports via the Internet. Researchers have developed a model for the city of Duisburg in Germany which combines data from road sensors with cellular automata simulations. The results show up-to-the-minute traffic flows throughout the city. (Check out their web site at http://traf2.uni-duisburg.de/OLSIM.

August 9 ♍
S M T W T F S

An equation for me has no meaning unless it expresses a thought of God.
            —Srinivasa Ramanujan

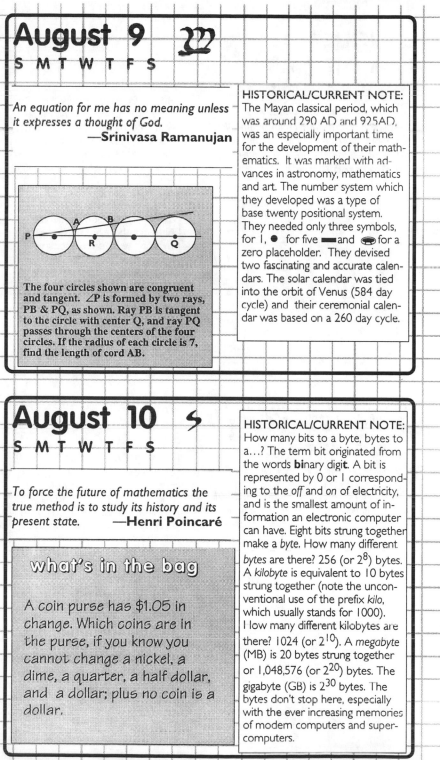

The four circles shown are congruent and tangent. ∠P is formed by two rays, PB & PQ, as shown. Ray PB is tangent to the circle with center Q, and ray PQ passes through the centers of the four circles. If the radius of each circle is 7, find the length of cord AB.

HISTORICAL/CURRENT NOTE:
The Mayan classical period, which was around 290 AD and 925AD, was an especially important time for the development of their mathematics. It was marked with advances in astronomy, mathematics and art. The number system which they developed was a type of base twenty positional system. They needed only three symbols, for 1, ● for five ▬and ⬭ for a zero placeholder. They devised two fascinating and accurate calendars. The solar calendar was tied into the orbit of Venus (584 day cycle) and their ceremonial calendar was based on a 260 day cycle.

August 10 ϟ
S M T W T F S

To force the future of mathematics the true method is to study its history and its present state.   **—Henri Poincaré**

what's in the bag

A coin purse has $1.05 in change. Which coins are in the purse, if you know you cannot change a nickel, a dime, a quarter, a half dollar, and a dollar; plus no coin is a dollar.

HISTORICAL/CURRENT NOTE:
How many bits to a byte, bytes to a...? The term bit originated from the words **bi**nary dig**it**. A bit is represented by 0 or 1 corresponding to the *off* and *on* of electricity, and is the smallest amount of information an electronic computer can have. Eight bits strung together make a *byte*. How many different *bytes* are there? 256 (or 2^8) bytes. A *kilobyte* is equivalent to 10 bytes strung together (note the unconventional use of the prefix *kilo*, which usually stands for 1000). How many different kilobytes are there? 1024 (or 2^{10}). A *megabyte* (MB) is 20 bytes strung together or 1,048,576 (or 2^{20}) bytes. The gigabyte (GB) is 2^{30} bytes. The bytes don't stop here, especially with the ever increasing memories of modern computers and supercomputers.

August 11
S M T W T F S

π's face was marked, and it was understood that none could behold it and live. But piercing eyes looked out from the mark, inexorable; cold, and enigmatic.
—**Bertrand Russell**

$$\left(\sqrt[5]{9}\right)^{3}\left(\sqrt[5]{3}\right)^{\sqrt{81}}$$

HISTORICAL/CURRENT NOTE:
It is worth noting that the Mayan solar calendar was more accurate than our Gregorian calendar. Their solar calendar was based on the Venus cycle, which made its Earth year 365.24 days. This produced an error of 1.98/10,000 of a day as compared to the 3.02/10,000 of a day for the Gregorian calendar. Although there are no records involving work on mathematical theory, their astronomy and mathematics were closely linked, as witness the accuracy of the astronomical calculations of two successive moons to be 29.5308642 days , while modern astronomical calculations give 29.53059 days.

August 12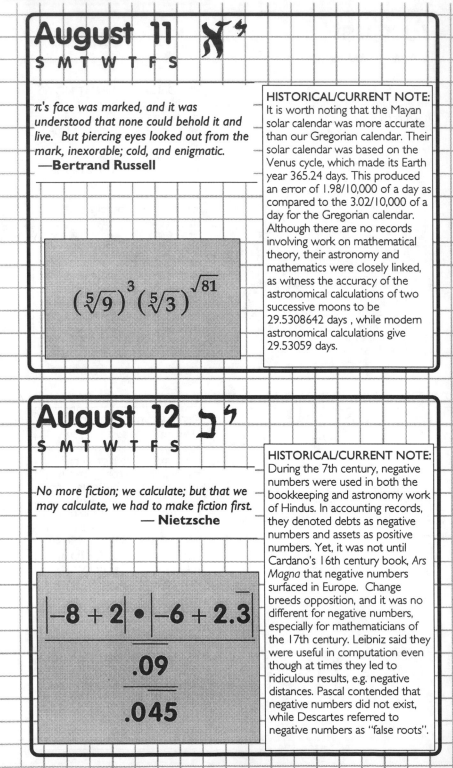
S M T W T F S

No more fiction; we calculate; but that we may calculate, we had to make fiction first.
— **Nietzsche**

$$\dfrac{\dfrac{\left|-8+2\right|\bullet\left|\overline{-6+2.3}\right|}{.09}}{.045}$$

HISTORICAL/CURRENT NOTE:
During the 7th century, negative numbers were used in both the bookkeeping and astronomy work of Hindus. In accounting records, they denoted debts as negative numbers and assets as positive numbers. Yet, it was not until Cardano's 16th century book, *Ars Magna* that negative numbers surfaced in Europe. Change breeds opposition, and it was no different for negative numbers, especially for mathematicians of the 17th century. Leibniz said they were useful in computation even though at times they led to ridiculous results, e.g. negative distances. Pascal contended that negative numbers did not exist, while Descartes referred to negative numbers as "false roots".

August 13
S M T W T F S

Music is the pleasure the human soul experiences from counting without being aware it is counting.
—**Gottfried Leibniz**

-3	-7	-5
4	-3	?
-12	21	-35
-8	18	-28

Figure out the pattern, and find the missing number

August 14
S M T W T F S

The profound study of nature is the most fertile source of mathematical discoveries.
—**Joseph Fourier**

simplify

$$\frac{6-2\left[3m+4(2m-1)\right]}{7-11m}$$

August 15 טו

S M T W T F S

The study of mathematics is apt to commence in disappointment....we are told that by its aid the stars are weighed and the billion of molecules in a drop of water are counted. Yet, like the ghost of Hamlet's father, this great science eludes the efforts of our mental weapons to grasp it.

—Alfred North Whitehead

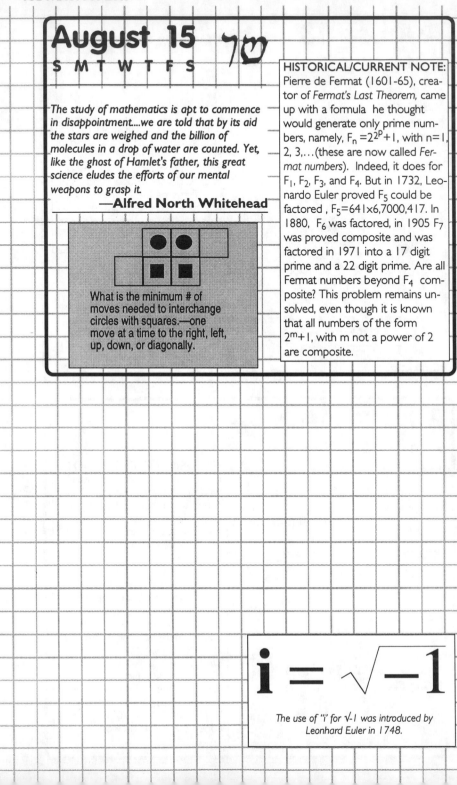

What is the minimum # of moves needed to interchange circles with squares.—one move at a time to the right, left, up, down, or diagonally.

HISTORICAL/CURRENT NOTE: Pierre de Fermat (1601-65), creator of *Fermat's Last Theorem*, came up with a formula he thought would generate only prime numbers, namely, $F_n = 2^{2^p} + 1$, with n=1, 2, 3,...(these are now called *Fermat numbers*). Indeed, it does for F_1, F_2, F_3, and F_4. But in 1732, Leonardo Euler proved F_5 could be factored, $F_5 = 641 \times 6,7000,417$. In 1880, F_6 was factored, in 1905 F_7 was proved composite and was factored in 1971 into a 17 digit prime and a 22 digit prime. Are all Fermat numbers beyond F_4 composite? This problem remains unsolved, even though it is known that all numbers of the form $2^m + 1$, with m not a power of 2 are composite.

$$i = \sqrt{-1}$$

The use of "i" for √-1 was introduced by Leonhard Euler in 1748.

August 16

S M T W T F S

In most sciences one generation tears down what another has built and what one has established another undoes. In mathematics alone each generation adds a new story to the old structure.
— **Hermann Hankle**

Find x, if x∈{ Integers }

$$\{5-3x<-10\} \cap \{4x+6<32\}$$

HISTORICAL/CURRENT NOTE:
Who is Dr. Itō? In 1942 Itō Kiyoshi developed and proved a theorem now called *Itō's Lemma*. The theorem gives a complicated mathematical equation* which describes a random behavior, and therefore can deal with chaotic systems. In the 1970s, the United States abandoned the gold standard and international currencies entered a floating rate system. As a result, prices of currencies, interest rates, stocks, bonds, futures, options could fluctuate dramatically from moment to moment. At the same time, in the world of mathematics, interest was emerging in the study of random behavior, which ended up applying to the analysis and prediction of fluctuating prices of financial products. Itō's mathematical equation directly applied to this evolving economics.

August 17

S M T W T F S

Logic is invincible because in order to combat logic it is necessary to use logic.
— **Pierre Butroux**

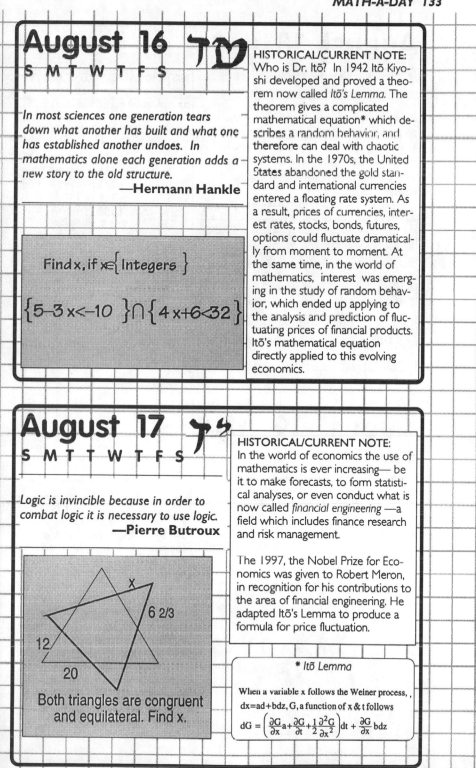

x

6 2/3

12

20

Both triangles are congruent and equilateral. Find x.

HISTORICAL/CURRENT NOTE:
In the world of economics the use of mathematics is ever increasing— be it to make forecasts, to form statistical analyses, or even conduct what is now called *financial engineering* —a field which includes finance research and risk management.

The 1997, the Nobel Prize for Economics was given to Robert Meron, in recognition for his contributions to the area of financial engineering. He adapted Itō's Lemma to produce a formula for price fluctuation.

* *Itō Lemma*

When a variable x follows the Weiner process, , dx=ad+bdz, G, a function of x & t follows

$$dG = \left(\frac{\partial G}{\partial x}a + \frac{\partial G}{\partial t} + \frac{1}{2}\frac{\partial^2 G}{\partial x^2}\right)dt + \frac{\partial G}{\partial x}bdz$$

August 18 ㄇ

S M T W T F S

In these days of conflict between ancient and modern studies, there must surely be something to be said for a study which did not begin with Pythagoras and will not end with Einstein, but is the oldest and youngest of all.

—G.H. Hardy

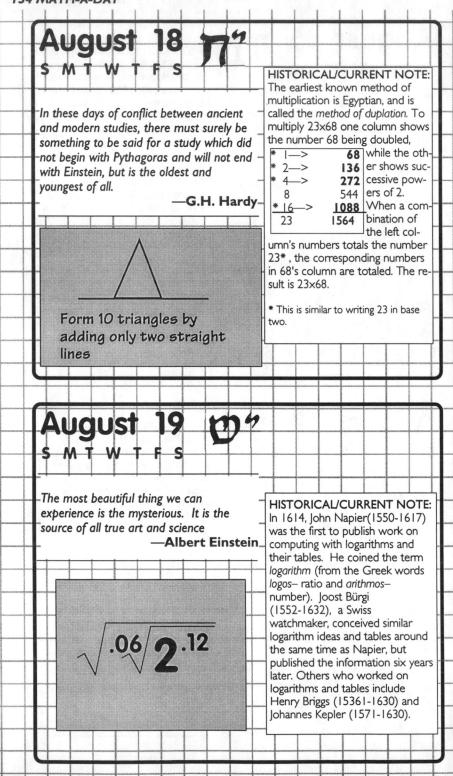

Form 10 triangles by adding only two straight lines

HISTORICAL/CURRENT NOTE:
The earliest known method of multiplication is Egyptian, and is called the *method of duplation*. To multiply 23x68 one column shows the number 68 being doubled,

*	1—>	68
*	2—>	136
*	4—>	272
	8	544
*	16—>	1088
	23	1564

while the other shows successive powers of 2. When a combination of the left column's numbers totals the number 23*, the corresponding numbers in 68's column are totaled. The result is 23x68.

* This is similar to writing 23 in base two.

August 19 ㄥ

S M T W T F S

The most beautiful thing we can experience is the mysterious. It is the source of all true art and science

—Albert Einstein

$$\sqrt[.06]{2^{.12}}$$

HISTORICAL/CURRENT NOTE:
In 1614, John Napier(1550-1617) was the first to publish work on computing with logarithms and their tables. He coined the term *logarithm* (from the Greek words *logos*– ratio and *arithmos*– number). Joost Bürgi (1552-1632), a Swiss watchmaker, conceived similar logarithm ideas and tables around the same time as Napier, but published the information six years later. Others who worked on logarithms and tables include Henry Briggs (15361-1630) and Johannes Kepler (1571-1630).

August 20
S M T W T F S

ב

There is nothing so easy but that it becomes difficult when you do it with reluctance. —**Terence**

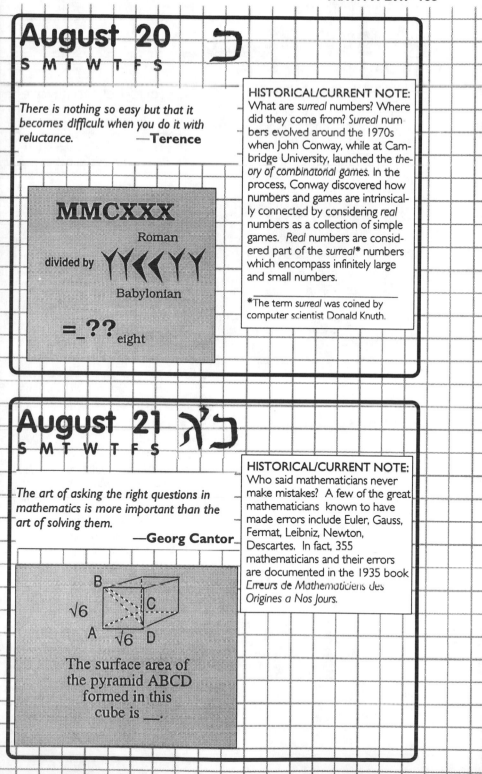

MMCXXX

Roman

divided by

Babylonian

$= \_??_{\text{eight}}$

HISTORICAL/CURRENT NOTE:
What are *surreal* numbers? Where did they come from? *Surreal* numbers evolved around the 1970s when John Conway, while at Cambridge University, launched the *theory of combinatorial games*. In the process, Conway discovered how numbers and games are intrinsically connected by considering *real* numbers as a collection of simple games. *Real* numbers are considered part of the *surreal** numbers which encompass infinitely large and small numbers.

---

*The term *surreal* was coined by computer scientist Donald Knuth.

August 21
S M T W T F S

כ'א

The art of asking the right questions in mathematics is more important than the art of solving them.

—**Georg Cantor**

HISTORICAL/CURRENT NOTE:
Who said mathematicians never make mistakes? A few of the great mathematicians known to have made errors include Euler, Gauss, Fermat, Leibniz, Newton, Descartes. In fact, 355 mathematicians and their errors are documented in the 1935 book *Erreurs de Mathematiciens des Origines a Nos Jours.*

$\sqrt{6}$

B

C

A $\sqrt{6}$ D

The surface area of the pyramid ABCD formed in this cube is __.

August 22 כב
S M T W T F S

Mechanics is the paradise of the mathematical sciences, because by means of it one comes to the fruits of mathematics. —**Leonardo da Vinci**

the hat puzzle

From a box containing 1 black and 2 tan hats, a hat is placed on Tim and Jim's heads while they are blindfolded. After removing the blindfolds, they are asked if they know what color they have on? Each could see the other's hat. Finally, Tim said he didn't know his color. At which point, Jim said he knew his. How did he know, and what color was it?

HISTORICAL/CURRENT NOTE:
Imagine a craft sailing into outer-space. The evolving nanotech world may make such a ship feasible. The sails would be about 100 atoms thick, and would not be propelled by wind, but sunlight. It would be able to maneuver itself throughout the far reaches of space. It is light that propels these sails. Considering light as particles, i.e. photons without mass, momentum is created when these particles bounce off (are reflected off) the shining sail, and the momentum created by these particles bouncing off the sail would propel the craft.

August 23 כג
S M T W T F S

....What immortal hand or eye Dare frame thy fearful symmetry? — **William Blake**

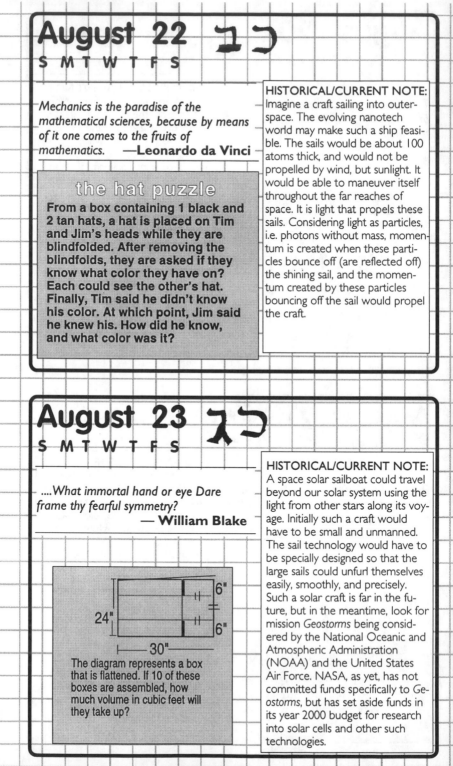

The diagram represents a box that is flattened. If 10 of these boxes are assembled, how much volume in cubic feet will they take up?

HISTORICAL/CURRENT NOTE:
A space solar sailboat could travel beyond our solar system using the light from other stars along its voyage. Initially such a craft would have to be small and unmanned. The sail technology would have to be specially designed so that the large sails could unfurl themselves easily, smoothly, and precisely. Such a solar craft is far in the future, but in the meantime, look for mission *Geostorms* being considered by the National Oceanic and Atmospheric Administration (NOAA) and the United States Air Force. NASA, as yet, has not committed funds specifically to *Geostorms*, but has set aside funds in its year 2000 budget for research into solar cells and other such technologies.

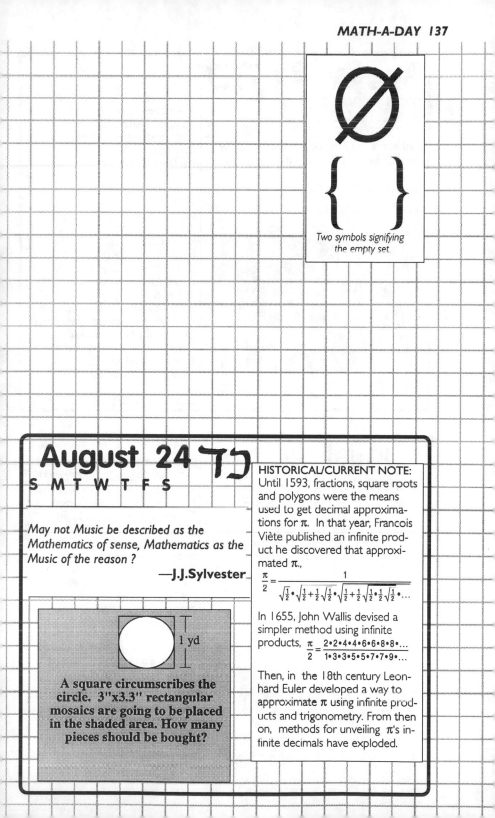

Two symbols signifying
the empty set.

August 24 ℼ

S M T W T F S

May not Music be described as the Mathematics of sense, Mathematics as the Music of the reason ?

—J.J.Sylvester

A square circumscribes the circle. 3"x3.3" rectangular mosaics are going to be placed in the shaded area. How many pieces should be bought?

1 yd

HISTORICAL/CURRENT NOTE:
Until 1593, fractions, square roots and polygons were the means used to get decimal approximations for π. In that year, Francois Viète published an infinite product he discovered that approximated π.,

$$\frac{\pi}{2} = \frac{1}{\sqrt{\frac{1}{2}} \cdot \sqrt{\frac{1}{2} + \frac{1}{2}\sqrt{\frac{1}{2}}} \cdot \sqrt{\frac{1}{2} + \frac{1}{2}\sqrt{\frac{1}{2} \cdot \frac{1}{2}\sqrt{\frac{1}{2}}}} \cdot \ldots}$$

In 1655, John Wallis devised a simpler method using infinite products, $\frac{\pi}{2} = \frac{2 \cdot 2 \cdot 4 \cdot 4 \cdot 6 \cdot 6 \cdot 8 \cdot 8 \cdot \ldots}{1 \cdot 3 \cdot 3 \cdot 5 \cdot 5 \cdot 7 \cdot 7 \cdot 9 \cdot \ldots}$

Then, in the 18th century Leonhard Euler developed a way to approximate π using infinite products and trigonometry. From then on, methods for unveiling π's infinite decimals have exploded.

August 25 כה

S M T W T F S

God made the integers, all the rest is the work of man. —**Leopold Kronecker**

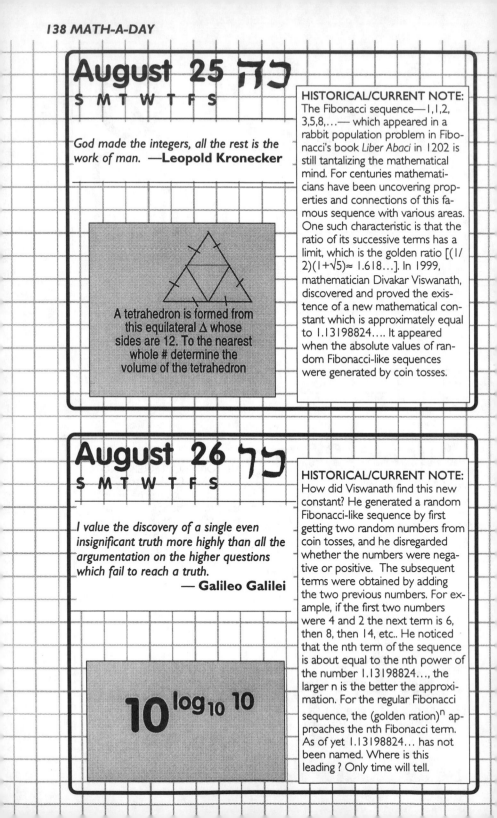

A tetrahedron is formed from this equilateral △ whose sides are 12. To the nearest whole # determine the volume of the tetrahedron

HISTORICAL/CURRENT NOTE: The Fibonacci sequence—1,1,2, 3,5,8,...— which appeared in a rabbit population problem in Fibonacci's book *Liber Abaci* in 1202 is still tantalizing the mathematical mind. For centuries mathematicians have been uncovering properties and connections of this famous sequence with various areas. One such characteristic is that the ratio of its successive terms has a limit, which is the golden ratio [$(1/2)(1+\sqrt{5})\approx 1.618...$]. In 1999, mathematician Divakar Viswanath, discovered and proved the existence of a new mathematical constant which is approximately equal to 1.13198824.... It appeared when the absolute values of random Fibonacci-like sequences were generated by coin tosses.

August 26 כו

S M T W T F S

I value the discovery of a single even insignificant truth more highly than all the argumentation on the higher questions which fail to reach a truth.
— **Galileo Galilei**

$$10^{\log_{10} 10}$$

HISTORICAL/CURRENT NOTE: How did Viswanath find this new constant? He generated a random Fibonacci-like sequence by first getting two random numbers from coin tosses, and he disregarded whether the numbers were negative or positive. The subsequent terms were obtained by adding the two previous numbers. For example, if the first two numbers were 4 and 2 the next term is 6, then 8, then 14, etc.. He noticed that the nth term of the sequence is about equal to the nth power of the number 1.13198824..., the larger n is the better the approximation. For the regular Fibonacci sequence, the (golden ration)^n approaches the nth Fibonacci term. As of yet 1.13198824... has not been named. Where is this leading ? Only time will tell.

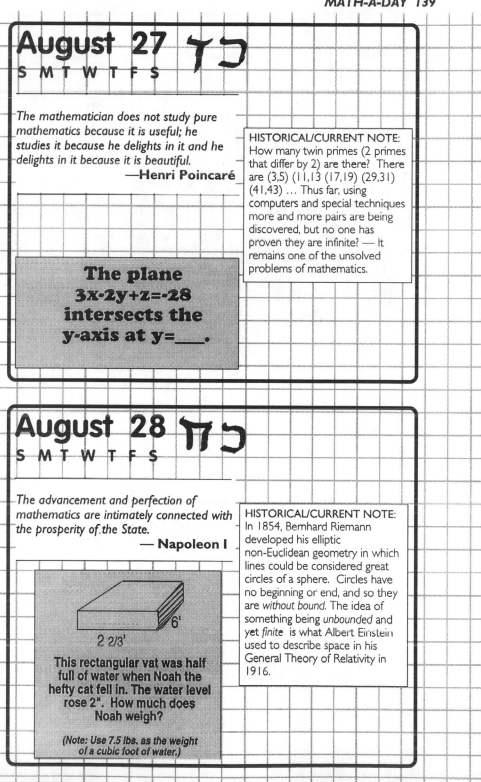

August 27 כז
S M T W T F S

The mathematician does not study pure mathematics because it is useful; he studies it because he delights in it and he delights in it because it is beautiful.
—**Henri Poincaré**

HISTORICAL/CURRENT NOTE: How many twin primes (2 primes that differ by 2) are there? There are (3,5) (11,13 (17,19) (29,31) (41,43) ... Thus far, using computers and special techniques more and more pairs are being discovered, but no one has proven they are infinite? — It remains one of the unsolved problems of mathematics.

The plane
3x-2y+z=-28
intersects the
y-axis at y=___.

August 28 כח
S M T W T F S

The advancement and perfection of mathematics are intimately connected with the prosperity of the State.
— **Napoleon I**

HISTORICAL/CURRENT NOTE: In 1854, Bernhard Riemann developed his elliptic non-Euclidean geometry in which lines could be considered great circles of a sphere. Circles have no beginning or end, and so they are *without bound*. The idea of something being *unbounded* and yet *finite* is what Albert Einstein used to describe space in his General Theory of Relativity in 1916.

6'

2 2/3'

This rectangular vat was half full of water when Noah the hefty cat fell in. The water level rose 2". How much does Noah weigh?

(Note: Use 7.5 lbs. as the weight of a cubic foot of water.)

August 29 ʊ⅂

S M T W T F S

Nature is pleased with simplicity, and affects not the pomp of superfluous causes. —**Isaac Newton**

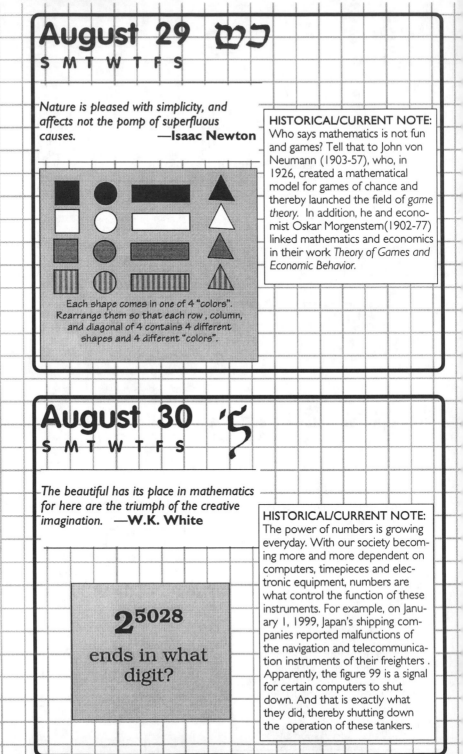

Each shape comes in one of 4 "colors". Rearrange them so that each row, column, and diagonal of 4 contains 4 different shapes and 4 different "colors".

HISTORICAL/CURRENT NOTE: Who says mathematics is not fun and games? Tell that to John von Neumann (1903-57), who, in 1926, created a mathematical model for games of chance and thereby launched the field of *game theory*. In addition, he and economist Oskar Morgenstern(1902-77) linked mathematics and economics in their work *Theory of Games and Economic Behavior*.

August 30 ⅃

S M T W T F S

The beautiful has its place in mathematics for here are the triumph of the creative imagination. —**W.K. White**

$$2^{5028}$$

ends in what digit?

HISTORICAL/CURRENT NOTE: The power of numbers is growing everyday. With our society becoming more and more dependent on computers, timepieces and electronic equipment, numbers are what control the function of these instruments. For example, on January 1, 1999, Japan's shipping companies reported malfunctions of the navigation and telecommunication instruments of their freighters . Apparently, the figure 99 is a signal for certain computers to shut down. And that is exactly what they did, thereby shutting down the operation of these tankers.

August 31

S M T W T F S

There is no sharply drawn line between those contradictions which occur in the daily work of every mathematician...and the major paradoxes which provide food for logical thought for decades and sometimes centuries.

—Nicholas Bourbaki

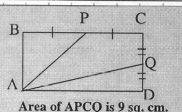

Area of APCQ is 9 sq. cm. Determine the area of rectangle ABCD.

HISTORICAL/CURRENT NOTE: Leonardo da Vinci emphasized the importance of understanding and studying mathematics as a framework to his various works and projects. In the late 15th century da Vinci drew a stallion which was to be a bronze sculpture towering 24 feet high. In the design of the horse he incorporated the golden rectangle's dimension to create dynamic symmetry. He then made a full sized clay model, which was destroyed by French troops invading Milan. Unfortunately, the sculpture was never cast. 500 years later, a foundation endowed by Charles Dent, had the sculpture recreated and two cast in bronze. One for a gift to Milan, which was unveiled on September 10, 1999, and the other for display in Beacon, Iowa.

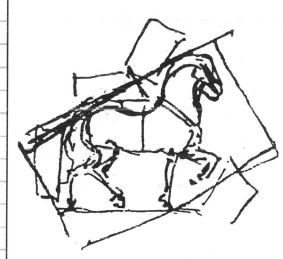

Leonardo da Vinci's sketch of his horse

September 1 —

S M T W T F S

September dates are written using the Brahmi numerals (c. 3rd century B.C.)

"I only took the regular course—the different branches of Arithmetic — Ambition, Distraction, Uglification and Derision."
— **Lewis Carroll** *Alice in Wonderland*

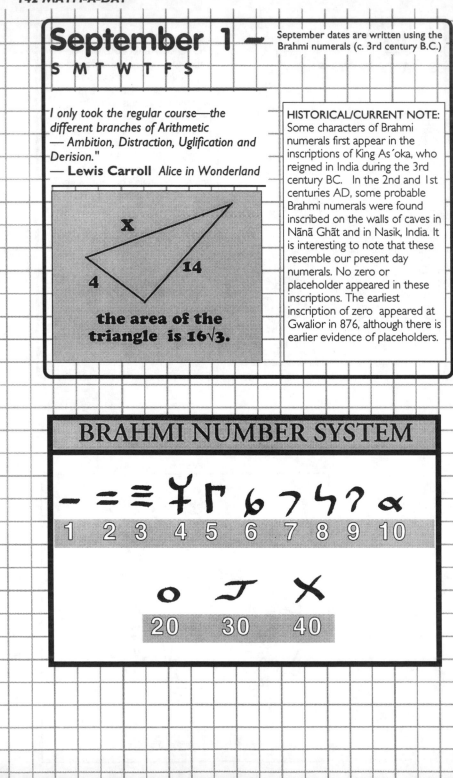

X

4

14

the area of the triangle is 16√3.

HISTORICAL/CURRENT NOTE: Some characters of Brahmi numerals first appear in the inscriptions of King As´oka, who reigned in India during the 3rd century BC. In the 2nd and 1st centuries AD, some probable Brahmi numerals were found inscribed on the walls of caves in Nānā Ghāt and in Nasik, India. It is interesting to note that these resemble our present day numerals. No zero or placeholder appeared in these inscriptions. The earliest inscription of zero appeared at Gwalior in 876, although there is earlier evidence of placeholders.

BRAHMI NUMBER SYSTEM

1	2	3	4	5	6	7	8	9	10

20	30	40

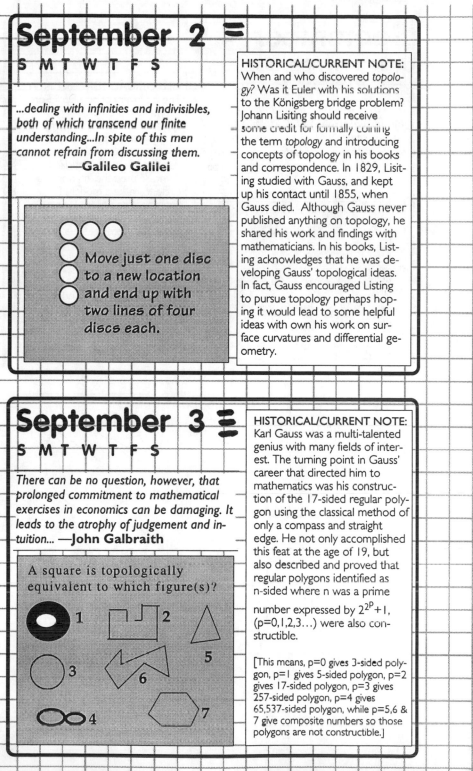

September 2

S M T W T F S

...dealing with infinities and indivisibles, both of which transcend our finite understanding...In spite of this men cannot refrain from discussing them.
—Galileo Galilei

Move just one disc to a new location and end up with two lines of four discs each.

HISTORICAL/CURRENT NOTE:
When and who discovered *topology*? Was it Euler with his solutions to the Königsberg bridge problem? Johann Lisiting should receive some credit for formally coining the term *topology* and introducing concepts of topology in his books and correspondence. In 1829, Listing studied with Gauss, and kept up his contact until 1855, when Gauss died. Although Gauss never published anything on topology, he shared his work and findings with mathematicians. In his books, Listing acknowledges that he was developing Gauss' topological ideas. In fact, Gauss encouraged Listing to pursue topology perhaps hoping it would lead to some helpful ideas with own his work on surface curvatures and differential geometry.

September 3

S M T W T F S

There can be no question, however, that prolonged commitment to mathematical exercises in economics can be damaging. It leads to the atrophy of judgement and intuition... —John Galbraith

A square is topologically equivalent to which figure(s)?

1 2 3 4 5 6 7

HISTORICAL/CURRENT NOTE:
Karl Gauss was a multi-talented genius with many fields of interest. The turning point in Gauss' career that directed him to mathematics was his construction of the 17-sided regular polygon using the classical method of only a compass and straight edge. He not only accomplished this feat at the age of 19, but also described and proved that regular polygons identified as n-sided where n was a prime number expressed by $2^{2^P}+1$, (p=0,1,2,3...) were also constructible.

[This means, p=0 gives 3-sided polygon, p=1 gives 5-sided polygon, p=2 gives 17-sided polygon, p=3 gives 257-sided polygon, p=4 gives 65,537-sided polygon, while p=5,6 & 7 give composite numbers so those polygons are not constructible.]

September 4 ♀

S M T W T F S

> *Only by taking an infinitesimally small unit for observation and attaining to the art of integrating them can we hope to arrive at the laws of history.*
>
> **—Tolstoy**
> *War and Peace*

Simplify

$$\left(\dfrac{\dfrac{4}{x^2} - \dfrac{4}{y^2}}{\dfrac{20}{x} - \dfrac{20}{y}} \right)\left(\dfrac{30xy}{x+y} \right)$$

HISTORICAL/CURRENT NOTE:
The discovery of non-Euclidean geometries prompted mathematicians to create models of non-Euclidean worlds that described these unusual geometries. Mathematician Henri Poincaré used the interior of a circle to make a model of hyperbolic geometry. *Suppose a point P is drawn off a given line L. In hyperbolic geometry there are infinitely many lines that can be drawn through P and parallel to line L.* With his model, lines are either represented by diameters or arcs whose ends are perpendicular to the circle.

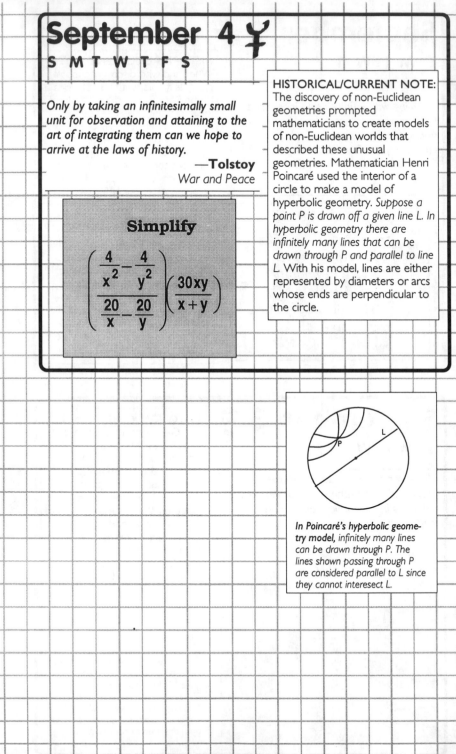

In Poincaré's hyperbolic geometry model, infinitely many lines can be drawn through P. The lines shown passing through P are considered parallel to L since they cannot interesect L.

September 5

S M T W T F S

I had been to school...and could say the multiplication tables up to 6 x 7 = 35, and don't reckon I could ever get any further than if I was to live forever. I don't take stock of mathematics.

—**Mark Twain**

$$11\frac{1}{4} + 2\frac{13}{16} + \frac{45}{64} + \frac{45}{256} + \cdots$$

HISTORICAL/CURRENT NOTE:
The place was India. The man Srinivasa Ramanujan (1887-1920), a poor Indian clerk who was so captivated by mathematics he avidly read any books he could get his hands on. Astonishingly, he taught himself mathematics and in the process he developed mathematical ideas he did not know already existed and came up with ingenious and novel ways to solve certain problems. He wrote to two British mathematicians, but neither ever replied. His third letter was to to G.H. Hardy. Hardy recognized Ramanujan's amazing talent, genius and drive, and invited him to Cambridge. Unfortunately, Ramanujan died at the age of 33 of tuberculosis. His papers and work have not been completely studied and understood, but his mathematics is already being applied to such areas as infinite series, elliptic functions, analytical number theory.

September 6

S M T W T F S

Can human reason without experience discover by pure thinking properties of real things?

—**Albert Einstein**

$$\frac{17}{1+\cfrac{1}{1-\frac{1}{9}}}$$

HISTORICAL/CURRENT NOTE:
What's PHARDO? It's the prototype for France's compact variation atomic fountain clock, designed for the International Space Station. Because of the space station's lack of gravity, the regular atomic fountain clock had to be modified. Since the atoms cannot cascade down without gravity, the adapted version has the atoms being propelled slowly across a microwave cavity.

September 7 7

S M T W T F S

Our knowledge of the first principles, such as space, time, motion, number, is as certain as any knowledge we obtain by reasoning. As a matter of fact, this knowledge provided by our hearts and instinct is necessarily the basis on which our reasoning has to build its conclusions.
—**Blaise Pascal**

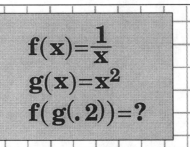

$$f(x) = \frac{1}{x}$$
$$g(x) = x^2$$
$$f(g(.2)) = ?$$

HISTORICAL/CURRENT NOTE:
Mathematician Heron (1st century AD) is best known for his technique of finding the area of a triangle when the lengths of its three sides are given*. Yet, Heron was an inventor and writer. He excelled in applied mathematics and demonstrated how to find the volumes of cones, prisms, pyramids and other figures. He explored optics and invented numerous mechanical devices which he described in his book *Pneumatica*.

* If a,b, and c represent the three sides of a triangle, then Heron's theorem gives its area as

$$\sqrt{s(s-a)(s-b)(s-c)} \quad \text{where} \quad s = \frac{a+b+c}{2}$$

September 8 ␄

S M T W T F S

The Analytical Engine has no pretensions whatever to originate anything. It can do whatever we know how to order it to perform. It can follow analysis, but it has no power of anticipating any analytical revelations or truths. Its province is to assist us in making available what we are already acquainted with. —**Ada Lovelace**
on Charles Babbage's computer

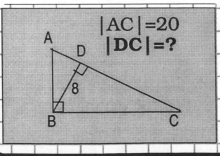

|AC| = 20
|DC| = ?

HISTORICAL/CURRENT NOTE:
The date May 11, 1997 marked a milestone in the study of AI (artificial intelligence). The game of chess was the testing ground. In 1985 Feng-hsiung Hsu, Thomas Anantharaman, and Murray Campbell developed the needed circuitry, programming and strategies for their machine, named Chiptest, to play chess. Over the years computers have participated in computer chess tournaments, but the first computer versus human contest took place in 1989 with Garry Kasparov playing against Deep Thought (a modified Chiptest). Kasparov won this challenge with ease. Deep Thought and its human team moved to IBM and evolved it to the computer Deep Blue, capable of testing 100-million positions a second. Again Kasparov agreed to another match in 1996, and again Kasparov won the match. The next rematch between Kasparov and Deep Blue (revved-up and with an expanded repertoire) was the fateful 7-game match of May 11, 1997—with 1 game for Kasparov, 3 for Deep Blue, and 3 draws.

September 9 ?

S M T W T F S

Beware gentle knight, there is no greater monster than reason. — **Cervantes**

AGE problem

**Harry is one year older than twice his sister's age. In three years he will be nine years older than his sister.
How old is Harry?**

HISTORICAL/CURRENT NOTE:
What happened to Deep Blue? When will the next match take place? Kasparov would probably be glad to reply. It's the winner who's in hiding. When you're ahead, why rock the boat, and that is exactly what's happened — IBM had Deep Blue dismantled. It seems that artifical intellience breeds artificial freedom.

September 10 ∝

S M T W T F S

..an irrational number...lies hidden in a cloud of infinity. —**Michael Stifel**

1,1,2,3,6,9, 15,135,150, 285,?,...

HISTORICAL/CURRENT NOTE:
Mathematics, mechanics and motion were hot scientific topics in the 17th century, which explains the keen interest of mathematicians in the cycloid at this time. The *cycloid is the curve traced by the path of a fixed point on a circle which rolls smoothly on a plane*. Galileo, Pascal, Roberval, Torrecelli, Descartes, Fermat, Wren, Wallis, Huygens, J. Bernoulli, Leibniz and Newton are among the famous mathematicians/scientists who feverishly sought to understand this fascinating curve. In the process, much hostility was created with arguments, accusations of plagiarism and about who discovered what first.

September 11 α-

S M T W T F S

...the feeling of mathematical beauty, of the harmony of numbers and of forms, of geometric elegance. It is a geninely esthetic feeling, which all mathematicians know. And this is sensitivity.

—**Henri Poicaré**

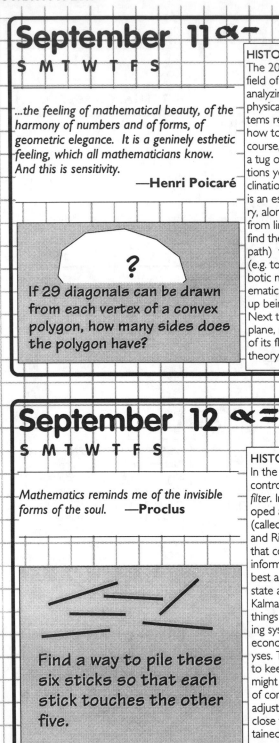

?

If 29 diagonals can be drawn from each vertex of a convex polygon, how many sides does the polygon have?

HISTORICAL/CURRENT NOTE:
The 20th century mathematical field of *control theory* encompasses analyzing what controls various physical systems, how these systems react to things around them, how to steer them along a desired course, and how to avoid creating a tug of war between the directions you want and the natural inclination. Mathematical modeling is an essential tool of control theory, along with the mathematics from linear algebra (which helps find the best or most economic path) to combinatorial analysis (e.g. to determine a range of robotic movements). Name a mathematical field, and it probably ends up being used in control theory. Next time you are up in an airplane, remember that the stability of its flight path relies on control theory.

September 12 α=

S M T W T F S

Mathematics reminds me of the invisible forms of the soul. —**Proclus**

Find a way to pile these six sticks so that each stick touches the other five.

HISTORICAL/CURRENT NOTE:
In the realm of linear algebra and control theory, we find the *Kalman filter*. In 1960 Rudolf Kalman developed a mathematical algorithm (called *Kalman filter* which Kalman and Richard Bucy later refined) that continually updates imprecise information to come up with the best approximation of a system's state at a particular moment. The Kalman filter works with such things as satellite and missile tracking systems, navigation of airplanes, economic and environmental analyses. The idea behind it is to try to keep ahead of changes that might send a nonlinear system out of control. Thus, by making minute adjustments along the way, a state close to equilibrium can be maintained.

September 13 ∝ ≡

S M T W T F S

..in the future more and more theoritical physics will command a deep knowledge of mathematical principals; and also that mathematicians will no longer limit themselves so exclusively to the aesthetic development of mathematical abstractions.
—George David Birkoff

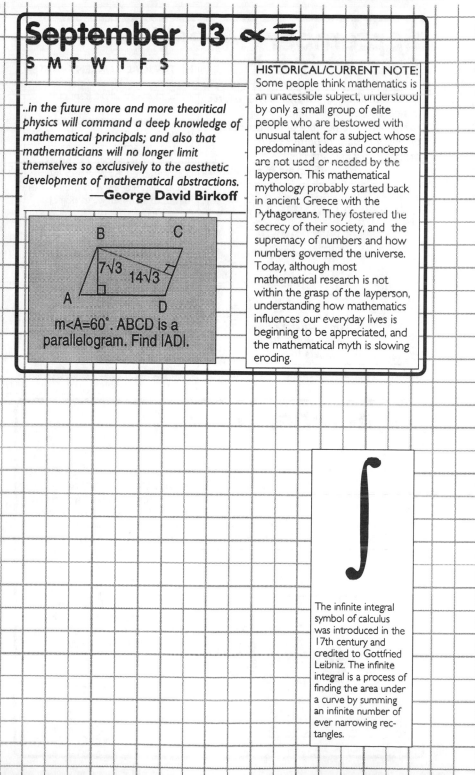

m<A=60°. ABCD is a parallelogram. Find |AD|.

HISTORICAL/CURRENT NOTE:
Some people think mathematics is an unacessible subject, understood by only a small group of elite people who are bestowed with unusual talent for a subject whose predominant ideas and concepts are not used or needed by the layperson. This mathematical mythology probably started back in ancient Greece with the Pythagoreans. They fostered the secrecy of their society, and the supremacy of numbers and how numbers governed the universe. Today, although most mathematical research is not within the grasp of the layperson, understanding how mathematics influences our everyday lives is beginning to be appreciated, and the mathematical myth is slowing eroding.

∫

The infinite integral symbol of calculus was introduced in the 17th century and credited to Gottfried Leibniz. The infinite integral is a process of finding the area under a curve by summing an infinite number of ever narrowing rectangles.

September 14 ∝Ƴ

S M T W T F S

Nature exhibits not simply a higher degree but an altogether different level of complexity. The number of distinct scales of length of natural patterns is for all practical purposes infinite.
—Benoit Mandelbrot

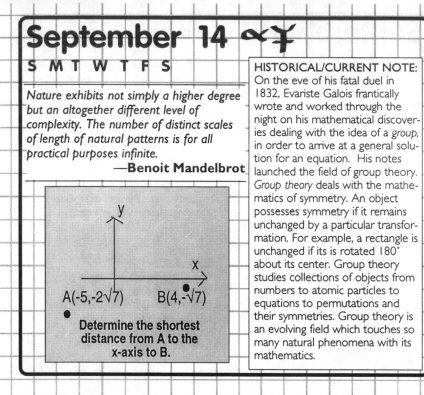

A(-5,-2√7)   B(4,-√7)

Determine the shortest distance from A to the x-axis to B.

HISTORICAL/CURRENT NOTE: On the eve of his fatal duel in 1832, Evariste Galois frantically wrote and worked through the night on his mathematical discoveries dealing with the idea of a *group*, in order to arrive at a general solution for an equation. His notes launched the field of group theory. *Group theory* deals with the mathematics of symmetry. An object possesses symmetry if it remains unchanged by a particular transformation. For example, a rectangle is unchanged if its is rotated 180° about its center. Group theory studies collections of objects from numbers to atomic particles to equations to permutations and their symmetries. Group theory is an evolving field which touches so many natural phenomena with its mathematics.

September 15 ∝Ⱦ

S M T W T F S

Cryptology is a science of deduction and controlled experiment; hypotheses are formed, tested and often discarded...The code "breaks"...when likely leads appear faster than they can be followed up.
—John Chadwick

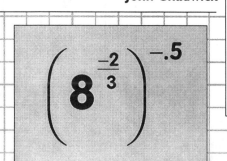

$$\left(8^{\frac{-2}{3}} \right)^{-.5}$$

HISTORICAL/CURRENT NOTE: Mathematicians and physicists are putting their heads together as well as their disciplines to tackle *quantum chaos*— a combination of chaos theory and quantum mechanics. Quantum mechanics is the theory of particles and waves developed in the 1920s to try to explain the behavior of the world of atoms, electrons, etc.. Stir in chaos theory of the 1980s— the mathematics of nonlinear systems— and presto, *quantum chaos* emerges.

September 16 ∝ 6

S M T W T F S

*We apprehend time only when we have
marked motion...we measure movement by
time, but also time by movement.*
                                    —Aristotle

The Barber Paradox

In a particular village in the
Alps, the barber shaves all
those in the village who do
not shave themselves.
Who shaves the barber?

HISTORICAL/CURRENT NOTE:
The humanoid Data, of *Star Trek–
The Next Generation,* mastered the
violin and classical music in a matter of seconds. Far fetched? Chris
Raphael has been working on programming a computer to accom-
pany a live soloist, which is able to
adapt to the soloist during the performance. The program has the
computer trying to learn the soloist's style by generating a statistical
model of the soloist's style— in
essence, Rapheal "trains" the
computer by playing the oboe,
while the computer "learns"
Rapheal's style.

September 17 ∝ 7

S M T W T F S

*Might is geometry; joined with art,
resistless.*   **—Euripides**

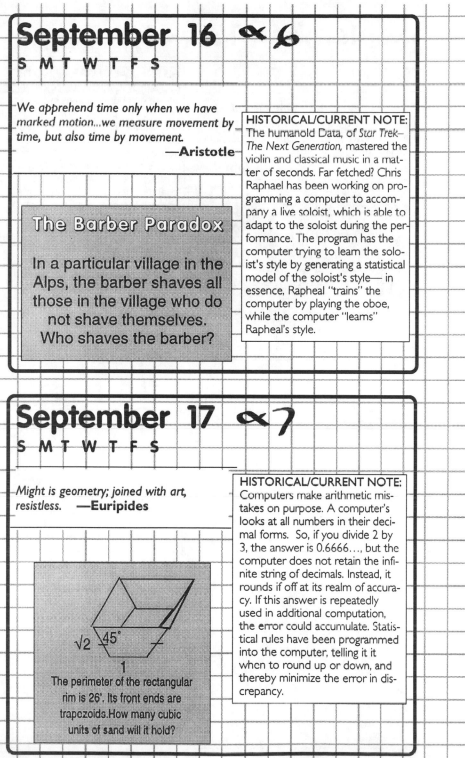

$\sqrt{2}$ 45°

1

The perimeter of the rectangular
rim is 26'. Its front ends are
trapezoids. How many cubic
units of sand will it hold?

HISTORICAL/CURRENT NOTE:
Computers make arithmetic mistakes on purpose. A computer's
looks at all numbers in their decimal forms. So, if you divide 2 by
3, the answer is 0.6666..., but the
computer does not retain the infinite string of decimals. Instead, it
rounds it off at its realm of accuracy. If this answer is repeatedly
used in additional computation,
the error could accumulate. Statistical rules have been programmed
into the computer, telling it
when to round up or down, and
thereby minimize the error in discrepancy.

September 18 ∝ ⅂

S M T W T F S

A marvelous neutrality have these things Mathematical, and also a strange partici-pation between things supernatural..., and things natural...

—**John Dee**

A H I ? O ?
U V W X Y

Discover what these letters have in common, and find the missing letters.

HISTORICAL/CURRENT NOTE:
Euclid proved the existence of an infinite number of primes in 300 BC, but who proved how many primes there are up to a particular number? *No one.* Karl Gauss first noticed that the number of prime numbers up to the number N was somehow related to the natural logarithm of N. In 1896, the *Prime Number Theorem*—which states there are *about* N/lnN prime numbers up to any number N chosen— was proven indepen-dently by mathematicians Jacques Hadamard of France and Charles-Jean de la Vallé Poussin of Belgium. Although this formula has about 5% error , it gives prime number hunters some added information for their quests.

September 19 ∝ ?

S M T W T F S

...regarding the fundamental investigations of mathematics, there is no final end-ing...no first beginning —**Felix Klein**

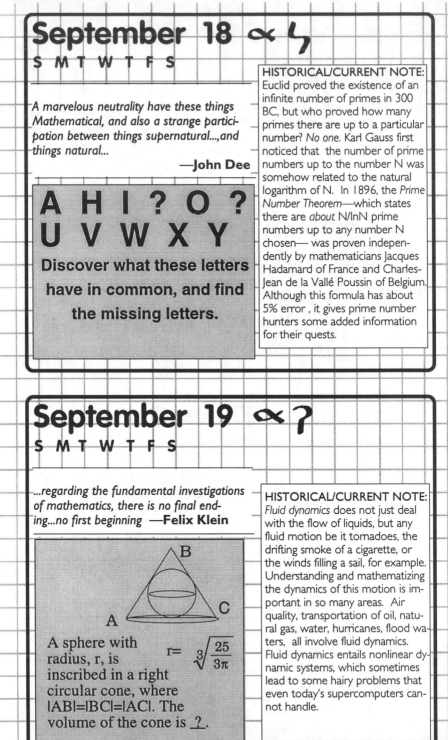

A sphere with radius, r, is inscribed in a right circular cone, where |AB|=|BC|=|AC|. The volume of the cone is _?_ .

$$r = \sqrt[3]{\frac{25}{3\pi}}$$

HISTORICAL/CURRENT NOTE:
Fluid dynamics does not just deal with the flow of liquids, but any fluid motion be it tornadoes, the drifting smoke of a cigarette, or the winds filling a sail, for example. Understanding and mathematizing the dynamics of this motion is im-portant in so many areas. Air quality, transportation of oil, natu-ral gas, water, hurricanes, flood wa-ters, all involve fluid dynamics. Fluid dynamics entails nonlinear dy-namic systems, which sometimes lead to some hairy problems that even today's supercomputers can-not handle.

September 20 ⭕

S M T W T F S

You say you got a real solution,
Well, you know,
We'd all love to see the plan.
—**John Lennon & Paul McCartney**

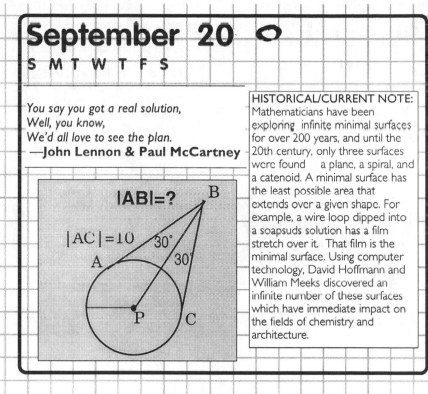

|AB|=?

|AC|=10

30°

30°

A

B

P

C

HISTORICAL/CURRENT NOTE:
Mathematicians have been exploring infinite minimal surfaces for over 200 years, and until the 20th century, only three surfaces were found — a plane, a spiral, and a catenoid. A minimal surface has the least possible area that extends over a given shape. For example, a wire loop dipped into a soapsuds solution has a film stretch over it. That film is the minimal surface. Using computer technology, David Hoffmann and William Meeks discovered an infinite number of these surfaces which have immediate impact on the fields of chemistry and architecture.

September 21 ⭕━

S M T W T F S

And who can doubt that it will lead to the worst disorders when minds created free by God are compelled to submit slavishly to an outside will? When we are told to deny our senses and subject them to the whim of others? When people devoid of whatsoever competence are made judges over experts and are granted authority to treat them as they please? These are the novelties which are apt to bring about the ruin of commonwealths and the subversion of the state. —**Galileo Galilei**

Shortcut ?

Without using a computer or calculator, add the numbers from 31 through 267.

HISTORICAL/CURRENT NOTE:
The geometry of nature, i.e. fractal geometry, has fascinating properties. Among its properties is *self-similarity*. This idea is considered to have originated in ancient times with Aristotle's viewing of the universe as a sequence of universes within universes. In the 17th century, Leibniz spoke of self-similarity of a line— a line made up of smaller lines. In the 20th century, self-similarity was brought into perspective and used by Benoit Mandelbrot in his work with fractals. Fractals generated from a basic pattern have clearly appearing self-similarity. This is especially evident when zeroing in reveals the repetition of a basic pattern, as with the equilateral triangle in the Koch snowflake. On the other hand, random fractals usually have statistical self-similarity.

September 22

S M T W T F S

I have discovered such wonderful things that I was amazed...out of nothing I have created a strange new universe.
— **Janos Bolyai**

WRONG
+ WRONG
RIGHT
If W=3,
what digit is
N?

HISTORICAL/CURRENT NOTE:
The ancients knew that the ratio of a circle's circumference to its diameter was a constant, which today we call π. Observation of patterns such as the appearance and reappearance of the constant π is a very important mathematical tool. In fact, mathematics is often described as the science of patterns. Mandelbrot discovered a pattern in fractals that led him to defining *fractal dimension*. He noticed that the ratio between a fractal's consecutive generated forms is constant, and this constant is referred to as that fractal's dimension. It was also discovered that the greater the dimension number, the more complex the fractal's curve.

September 23

S M T W T F S

Mountains are not cones, clouds are not spheres, trees are not cylinders, neither does lightning travel in a straight line. Almost everything around us is non-Euclidean. — **Benoit Mandelbrot**

Two boats are traveling toward each other at 9 mph and 5 mph. They are sending a signal back and forth at a rate of 14 miles per minute. After the signal has traveled 1680 miles, the boats meet. How far apart were the boats when the signal started?

HISTORICAL/CURRENT NOTE:
In 1977 a ciphertext called RSA-129 was created by Ronald Rivest, Adi Shamir, and Leonard Adleman. The RSA team put this out as a challenge to mathematical cryptographers, and had estimated that with existing computer technology, it would take over 20,000 years to decipher. In 1994 a loosely formed international group of over 600 code-breaker enthusiasts solved the cipher in about eight months. The decoded cipher text said *"The magic words are squeamish ossifrage."* The key to the numeric code was hidden in the prime factors of a 129-digit number.

Some mathematicians used this reversed D, to symbolize division in the 18th century.

September 24 o♀

S M T W T F S

The charm (of mathematics) lies chiefly...in the absolute certainty of its results; for that is what, beyond all mental treasures, the human intellect craves for. Let us be sure of something! More light, more light!
—**Charles Dodgson** (Lewis Carroll)

$$y=2\tan 10\theta$$

What is this trigonometric function's period in degrees?

HISTORICAL/CURRENT NOTE:
We often take for granted how easily the brain recognizes and compares objects. The Center for Intelligent Control Systems (CICS) is studying such things as pattern recognition and its connection to autonomous decision making. Breakthroughs in this area of computer science would mean computers would be able to read and compare medical scans for patients or track and identify radar readings for several crafts simultaneously. A lot more power, memory and data is needed before computers will be able to "see", "recognize", "compare" and "determine" the optimum solution.

September 25

S M T W T F S **or**

If I were to awaken after having slept for a thousand years, my first question would be: Has the Riemann hypothesis been proven? —**David Hilbert**

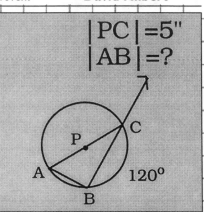

$|PC|=5"$
$|AB|=?$

$120°$

HISTORICAL/CURRENT NOTE: The *Riemann hypothesis* is one of the famous unsolved problems. During his life, Bernhard Riemann (1826-66) revealed some fascinating mathematical gems. Among these he discovered and proved a non-Euclidean geometry, introduced the world to worm holes and presented an ingenious hypothesis that has yet to be proven— the *Riemann hypothesis*. It is connected to what is called the zeta function—

$$\zeta(s)=1^s+(1/2)^s+(1/3)^s+\ldots=0.$$
and s is a complex number.

His hypothesis claims that the real part of the complex number s is always 1/2 whenever $\zeta(s)=0$. Although this has yet to be proven, much mathematics has been developed under the assumption it is true, especially in the area of number theory and more specifically prime numbers.

September 26 o 6

S M T W T F S

It would only be possible to imagine life or beauty as being "strictly mathematical" if we ourselves were such infinitely capable mathematicians as to be able to formulate their characteristics in mathematics so extremely complex that we have never yet invented them.
—**Theodore Andrea Cook**

HISTORICAL/CURRENT NOTE: The *Riemann hypothesis* does not apply only to mathematics. Physicists believe it has a direct connection to *quantum chaos*. Moreover, the zero solutions of the zeta function, i.e. $\zeta(s)=0$, can be considered energy levels of particles in quantum chaos. Therefore, quite a bit of physics is riding on the proof of the *Riemann hypothesis*.

$$32^{0.4}$$

September 27 o7

S M T W T F S

The laws of mathematics are not merely human inventions or creations. They simply "are"; they exist quite independently of the human intellect. The most that any man with a keen intellect can do is to find out that they are there and to take cognizance of them. **—M.C. Escher**

$$\frac{2\sqrt{1500x-22500}}{x} = 10$$

HISTORICAL/CURRENT NOTE:
In 1834, John Scott Russell recorded the first sighting of a soliton wave— a wave that retains its shape, size, and momentum. Today, the study of soliton waves and their properties is a very current topic. The mathematics developed to describe the soliton by the end of the 19th century, along with recent findings of *compacton* and *peakon equations*, explores the use of these waves to pack electronic data and send it efficiently along the information highway.

September 28 o5

S M T W T F S

It is a capital mistake to theorize before one has data.
—Sir Arthur Conan Doyle

$$5x^2 + \left(\sqrt{2}-60\right)x$$

$$=12\sqrt{2}$$

and x is an Integer.

HISTORICAL/CURRENT NOTE:
In 1976, *public key codes*— a method of encrypting information— was introduced by Whitfield Diffie and Martin Hellman of Stanford University. Public key cryptography allows complete strangers to communicate or carry on business transactions via the Internet. The sender's electronic data uses a secret mathematical "key" to encrypt a message and the receiver has a different mathematically related "key" to decipher it. In 1977, the practical implementation of public key codes was carried out by the work of Ronald Revest, Adi Shami and Leonard Adlemann.

September 29 o?

S M T W T F S

Oh these mathematicians make me tired! When you ask them to work out a sum they take a piece of paper, cover it with rows of A's B's and X's Y's...scatter a mess of flyspecks over them, and then give you an answer that's all wrong. —**Thomas Edison**

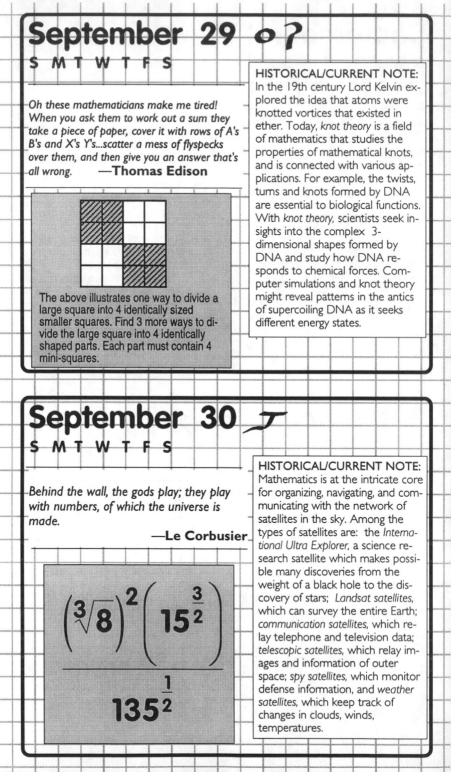

The above illustrates one way to divide a large square into 4 identically sized smaller squares. Find 3 more ways to divide the large square into 4 identically shaped parts. Each part must contain 4 mini-squares.

HISTORICAL/CURRENT NOTE:
In the 19th century Lord Kelvin explored the idea that atoms were knotted vortices that existed in ether. Today, *knot theory* is a field of mathematics that studies the properties of mathematical knots, and is connected with various applications. For example, the twists, turns and knots formed by DNA are essential to biological functions. With *knot theory*, scientists seek insights into the complex 3-dimensional shapes formed by DNA and study how DNA responds to chemical forces. Computer simulations and knot theory might reveal patterns in the antics of supercoiling DNA as it seeks different energy states.

September 30 ⌐

S M T W T F S

Behind the wall, the gods play; they play with numbers, of which the universe is made.

—**Le Corbusier**

$$\frac{\left(\sqrt[3]{8}\right)^2 \left(15^{\frac{3}{2}}\right)}{135^{\frac{1}{2}}}$$

HISTORICAL/CURRENT NOTE:
Mathematics is at the intricate core for organizing, navigating, and communicating with the network of satellites in the sky. Among the types of satellites are: the *International Ultra Explorer*, a science research satellite which makes possible many discoveries from the weight of a black hole to the discovery of stars; *Landsat satellites,* which can survey the entire Earth; *communication satellites,* which relay telephone and television data; *telescopic satellites,* which relay images and information of outer space; *spy satellites,* which monitor defense information, and *weather satellites,* which keep track of changes in clouds, winds, temperatures.

October 1

S M T W T F S

October dates are in the
Babylonian number system

*...for no human inquiry can be called
science unless it pursues its path through
mathematical exposition and
demonstration.* —**Leonardo da Vinci**

HISTORICAL/CURRENT NOTE:
Today's copyright laws do not
protect a book's title. The same
was true in ancient times, when
(circa 430 BC) the Greek mathematician Hippocrates of Chios
wrote his book *Elements*. Unfortunately, his work was lost,
but the same title was used by
Euclid in 300 BC for his famous
book *Elements*.

Tim, Tom, and Ted each use different
modes of transportation. One bikes, one
drives a car, and one roller blades. The
youngest, the roller bladder, is Tom's
best friend. Ted is older than the car
driver. Who uses what?

BABYLONIAN NUMERALS

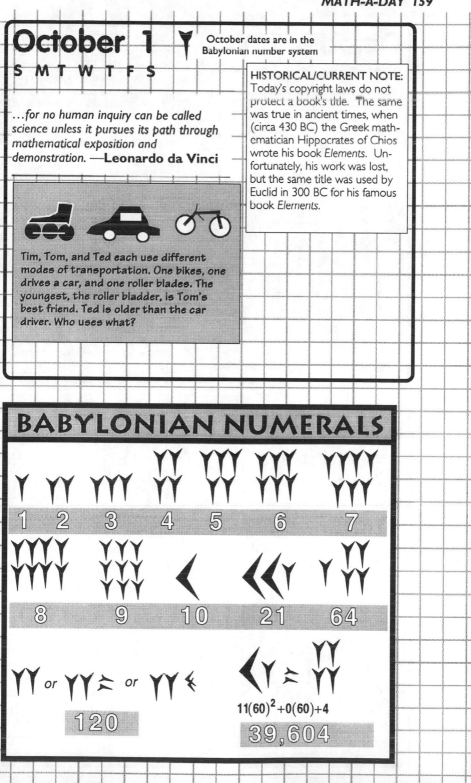

1	2	3	4	5	6	7

8	9	10	21	64

120

$11(60)^2 + 0(60) + 4$

39,604

October 2 ΥΥ
S M T W T F S

...a number is merely the product of our mind. —**Karl Gauss**

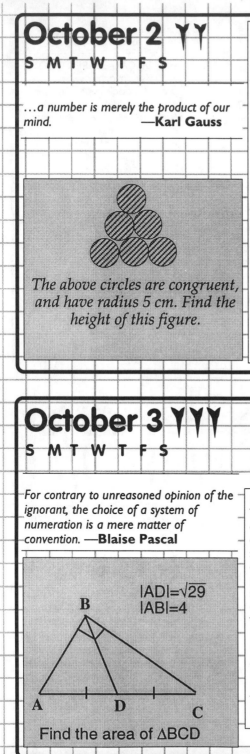

The above circles are congruent, and have radius 5 cm. Find the height of this figure.

HISTORICAL/CURRENT NOTE:
Imagine an object that is not 1, 2, or 3 dimensional, but some fractional dimension. Imagine an object that has finite area but infinite perimeter. Imagine a curve that is continuous everywhere but can have no tangents. Strange?
In the late 19th century, when these first surfaced in the works of Georg Cantor, Helge von Koch, Guiseppe Peano, conservative mathematicians labeled such objects *mathematical monsters* or anomalies. In the early 1900s other mathematicians, among them were Gaston Julia and Pierre Fatou, made contributions in this "mysterious" new area of mathematics with its unusual creatures. These mathematical monsters really came to life with the advent of modern computers and the1952-75 work of Benoit Mandelbrot, who named them *fractals*.

October 3 ΥΥΥ
S M T W T F S

For contrary to unreasoned opinion of the ignorant, the choice of a system of numeration is a mere matter of convention. —**Blaise Pascal**

|ADI=√29
|ABI=4

B

A   D   C

Find the area of ΔBCD

HISTORICAL/CURRENT NOTE:
Today, fractals can be used to describe the unusual shapes that don't conform to the mold of squares, circles, triangles, cones, etc.. Look at a cloud, a bag of popcorn, the shape of roots, the structure of a tree, the jaggedness of a rock. All these and any object that grows or changes can probably be captured by a fractal description. That's why they are readily used in creating sets for cinematography, in describing the changes in economics, in helping in the diagnosis of osteoporosis, — in short, they are almost everywhere.

October 4

S M T W T F S

Mathematics seems to endow one with something like a new sense.
—Charles Darwin

At a particular time of day the tree's shadow is 36 2/3', while a yardstick at this time has a shadow of (12/3) yard. How tall is the tree?

HISTORICAL/CURRENT NOTE: Nanoworld — the world of minute creatures. Imagine a robot so small it is invisible to the naked eye, or a computer the size of a bacterium. The mathematical foundations for nanoworld and nanotechnologies have taken centuries to evolve. From ancient times mathematics has been used to deal with the very small and the very large— for example, Zeno's paradoxes dealt with the infinitely small and large, Cantor's transfinite numbers described infinite sets of varying size, while Mandelbrot fractals deal with their infinite worlds within infinite worlds.

$$\sqrt[3]{}$$

Today's cube root symbol originated in France in the 17th century.

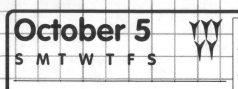

October 5
S M T W T F S

…It is truth very certain, that when it is not in our power to determine what is true, we ought to follow what is most probable. —**René Descartes**

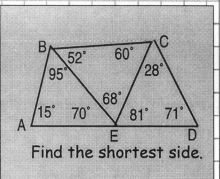

Find the shortest side.

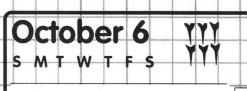

October 6
S M T W T F S

In any particular theory there is only as much real science as there is mathematics. —**Immanuel Kant**

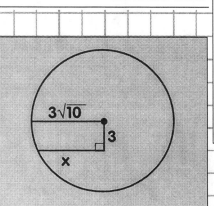

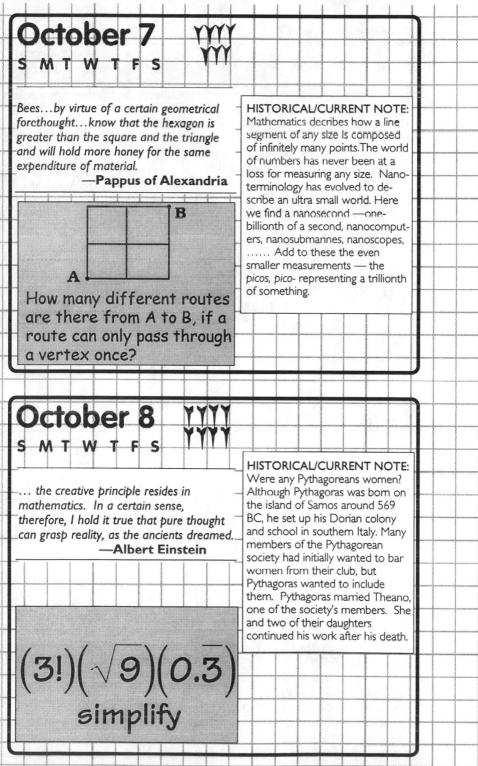

October 7

S M T W T F S

> Bees...by virtue of a certain geometrical forethought...know that the hexagon is greater than the square and the triangle and will hold more honey for the same expenditure of material.
> —**Pappus of Alexandria**

How many different routes are there from A to B, if a route can only pass through a vertex once?

HISTORICAL/CURRENT NOTE:
Mathematics describes how a line segment of any size is composed of infinitely many points. The world of numbers has never been at a loss for measuring any size. Nano-terminology has evolved to describe an ultra small world. Here we find a nanosecond —one-billionth of a second, nanocomputers, nanosubmarines, nanoscopes, ... Add to these the even smaller measurements — the *picos, pico-* representing a trillionth of something.

October 8

S M T W T F S

> ... the creative principle resides in mathematics. In a certain sense, therefore, I hold it true that pure thought can grasp reality, as the ancients dreamed.
> —**Albert Einstein**

HISTORICAL/CURRENT NOTE:
Were any Pythagoreans women? Although Pythagoras was born on the island of Samos around 569 BC, he set up his Dorian colony and school in southern Italy. Many members of the Pythagorean society had initially wanted to bar women from their club, but Pythagoras wanted to include them. Pythagoras married Theano, one of the society's members. She and two of their daughters continued his work after his death.

$$(3!)\left(\sqrt{9}\right)\left(0.\overline{3}\right)$$

simplify

October 9
S M T W T F S

Mathematics is often defined as the science of space and number...not until the recent resonance of computers and mathematics that a more apt definition became fully evident: mathematics is the science of patterns.

—Lynn Arthur Steen

$$\left(\sqrt[4]{\sqrt[3]{\sqrt[2]{64}}} \right)^8$$

HISTORICAL/CURRENT NOTE:
Although the Pythagorean theorem carries the name of the Greek mathematician Pythagoras, it was first discovered by a Mesopotamian mathematician around 1600 BC. But it was the Pythagoreans who first proved the theorem. The usefulness and fascination for this theorem has not diminished over the centuries, and has appeared in the mathematics of many civilizations. Probably more proofs of this theorem have been devised than any other theorem. The earliest existing proof is Euclid's *Forty-Seventh Proposition*. In addition to Euclid's proof of the Pythagorean theorem, others include— a paperfolding version, a multicolored proof by Oliver Byrne in the 1800s, Leonardo da Vinci's proof in the 15th century, one by President James Garfield.

October 10
S M T W T F S

Ten...this number was of old held high in honor, for such is the number of fingers by which we count. **—Ovid**

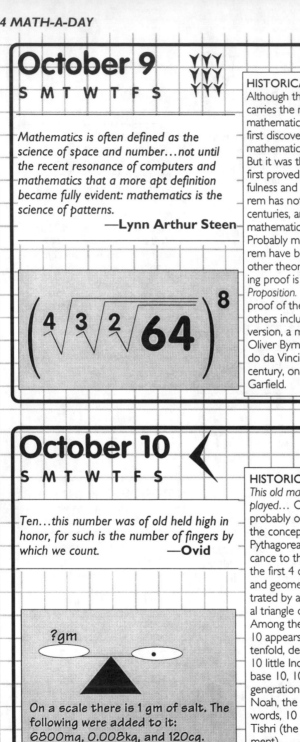

?gm

On a scale there is 1 gm of salt. The following were added to it: 6800mg, 0.008kg, and 120cg. How many grams does the scale now weigh?

HISTORICAL/CURRENT NOTE:
This old man he played 10, he played... Our 10 fingers were probably our first introduction to the concept of 10 things. The Pythagoreans attached a significance to the fact that the sum of the first 4 counting numbers is 10, and geometrically it could be illustrated by an equilateral triangle of dots. Among the places 10 appears are: tenfold, decade, 10 little Indians, base 10, 10 Commandments, 10 generations between Adam and Noah, the world was created in 10 words, 10 plagues on Egypt, 10 Tishri (the Jewish day of Atonement).

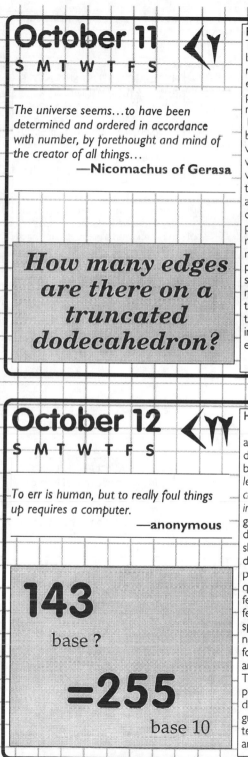

October 11

S M T W T F S

*The universe seems...to have been
determined and ordered in accordance
with number, by forethought and mind of
the creator of all things...*
—**Nicomachus of Gerasa**

*How many edges
are there on a
truncated
dodecahedron?*

HISTORICAL/CURRENT NOTE:
Today's computer viruses seem to
be generated as quickly, if not
more quickly, than biological viruses. Estimates claim 300 new computer viruses are created every
month. For example, in April of
1999, the Melissa virus startled
businesses, organizations, and individuals by the speed at which it
was spread. It entered computers
via e-mail, and invaded the electronic mail service of businesses,
agencies, organizations. Melissa's
designer chose a widely used computer software to conceal the virus, which may account for the
rapid spread of Melissa. The virus
posed as a *macro* in this popular
software. Once a letter was e-
mailed using the infected program,
the unsuspecting recipient opening
the letter activated the virus which
immediately searched the computer for an organizer program

continued on Oct. 12

October 12

S M T W T F S

*To err is human, but to really foul things
up requires a computer.*
—**anonymous**

143
base ?

=255
base 10

HISTORICAL/CURRENT NOTE:
and mailed itself to the first 50 addresses in the recipient's address
book. *Each one of these infected
letters was potentially able to repli-
cate itself 50 more times until this
incredible electronic chain letter*
grew so quickly that it overbur-
dened and forced e-mail sites to
shut down. Fortunately, Melissa
did not destroy data, but what's to
preclude a data killing virus? As
quickly as Melissa exploded, its in-
fections quickly subsided after a
few days because of the rapid response by virus monitoring businesses. They immediately informed the public of the danger,
and quickly developed vaccines.
The best protection against computer viruses is to frequently up-
date one's virus protection programs, encourage improved
technology of program designs,
and keep informed.

October 13 ⟨ＹＹＹ

S M T W T F S

We must admit with humility that, while number is purely a product of our minds, space has a reality outside our minds, so that we cannot completely prescribe its properties a priori. —**Karl Gauss**

Find the volume of this frustum.

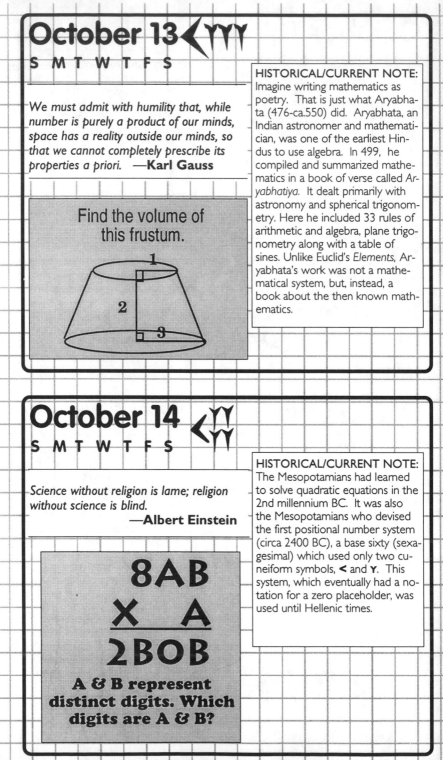

HISTORICAL/CURRENT NOTE:
Imagine writing mathematics as poetry. That is just what Aryabhata (476-ca.550) did. Aryabhata, an Indian astronomer and mathematician, was one of the earliest Hindus to use algebra. In 499, he compiled and summarized mathematics in a book of verse called *Aryabhatiya*. It dealt primarily with astronomy and spherical trigonometry. Here he included 33 rules of arithmetic and algebra, plane trigonometry along with a table of sines. Unlike Euclid's *Elements*, Aryabhata's work was not a mathematical system, but, instead, a book about the then known mathematics.

October 14 ⟨ＹＹ

S M T W T F S

Science without religion is lame; religion without science is blind.
—**Albert Einstein**

$$\begin{array}{r} 8AB \\ \times \quad A \\ \hline 2BOB \end{array}$$

A & B represent distinct digits. Which digits are A & B?

HISTORICAL/CURRENT NOTE:
The Mesopotamians had learned to solve quadratic equations in the 2nd millennium BC. It was also the Mesopotamians who devised the first positional number system (circa 2400 BC), a base sixty (sexagesimal) which used only two cuneiform symbols, **<** and **Ｙ**. This system, which eventually had a notation for a zero placeholder, was used until Hellenic times.

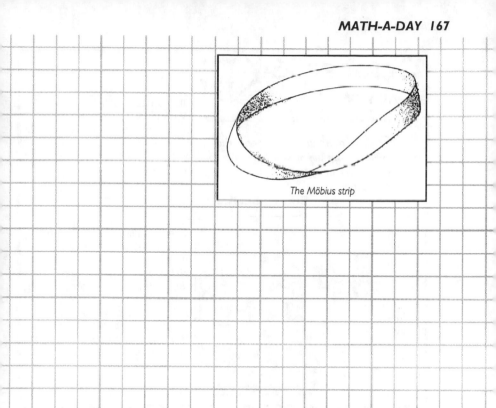

The Möbius strip

October 15
S M T T W T F S

...without (geometry)which no one can either be or become an absolute artist; but the blame for this should be laid upon their masters, who are themselves ignorant of this art. **—Albrecht Dürer**

15x15 magic square

Suppose a 15x15 magic square is made from the first 225 natural numbers. Without making the square, what will its magic number be?

HISTORICAL/CURRENT NOTE: Who would believe that gambling would lead to the discovery of mathematical ideas. Yet, major headway in *the theory of probability* began in 1654 when gambler Chevalier de Méré asked Blaise Pascal for help in *determining how the stakes should be divided when a game of chance is forced to stop prematurely.* Pascal wrote Pierre de Fermat about the problem, and the two of them through a series of letters launched the theory of probability. The first book on probability, *Liber de ludo aleae* (Book on the games of chance) was written by Girolamo Cardano(1501-76) and published post-humously in 1663. It was Pierre Laplace who defined probability as a ratio. Jacob Bernoulli(1654-1705) compiled works on probability and extended its ideas in his book *Ars Conjectando*, which was published posthumously in 1713.

October 16

S M T W T F S

The science of mathematics presents the most brilliant example of how pure reason may successfully enlarge its domain without the aid of experience .
—Emmanuel Kant

$$36^{2x-8} = 6^{3x}$$

HISTORICAL/CURRENT NOTE:
Was the Möbius strip first discovered by mathematician August Möbius(1790-1868)? It is unclear whether his should be the sole name on this fascinating one-sided topological object. Johann B. Lisitng's (1808-82) book *The Census of spatial complexes or generalizations of Euler's theorem on polyhedra* was published in 1861, and contains a description of the Möbius strip. Möbius did not publish his work about the strip until 1865. However, it was mentioned in 1858 in unpublished papers of both men. It seems that the idea occurred to both men independently, and they were both familiar with the work of Karl Gauss on topological ideas.

October 17

S M T W T F S

Reason's last step is the recognition that there are an infinite number of things which are beyond it. **—Blaise Pascal**

$$\frac{\dfrac{3}{5}}{a+1} = 9$$

HISTORICAL/CURRENT NOTE:
What is behind *smart* technology, machines and appliances? *Fuzzy logic.* The logic of yes, no, and maybe. This is not clear cut logic. Things are not just black or white, but can be in a gray area. Here we find an automobile being designed to learn the driving habits of its owner, so that it can anticipate accidents and take preventive action. Traditional computers rely on the binary number system (0s and 1s) to run the hardware and design their programs. In the future, fuzzy logic and fuzzy numbers (numbers which are not a specific amount will be important). For example, a fuzzy 1 could be defined as any quantity between 0.5 and 1, that is, a range of numbers would be represented by a fuzzy 1. The future of new technologies will depend on fuzzy stuff.

October 18

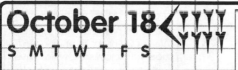

S M T W T F S

Nothing has afforded me so convincing a proof of the unity of the Deity as these purely mental conceptions of numerical and mathematical science ...
—**Mary Somerville**

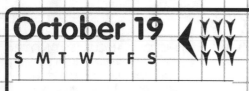

A boat heads out of the harbor at a bearing of 60°. After 14 miles it changes its bearing to 200°. After 3 miles on this route, what is its distance from the harbor?

HISTORICAL/CURRENT NOTE: Algebra and Greek mathematics are not usually linked together. Greek mathematics was mainly concerned with uncovering geometric truths and proving them. The development of a logical progression of mathematical ideas was one of the major contributions of the Greeks. Yet, *algebraic problems were not alien to the Greeks; it is just that their approach to solving them relied heavily on geometric means.* For example, we find that Euclid solved equations of the form $x^2+bx=b$ and $x^2+bx=c^2$ by completing the geometric square and ignoring negative answers. Similar algebraic examples are found in the works of Heron and Hippocrates. Diophantus dealt mainly with algebra and indeterminate algebraic problems and equations, which have come to be called *Diophantine equations.*

October 19

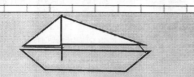

S M T W T F S

Fourier is a mathematical poem.
—**Lord Kelvin**

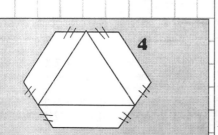

This is a regular hexagon. Determine the length of the sides of the triangle.

HISTORICAL/CURRENT NOTE: *Who will unravel the mystery of the constant γ?* What kind of number is it —rational, irrational, transcendental? As yet nobody knows, even though it pops up in math problems. What we know for sure is that the average number of divisors of all numbers from 1 to n is very close to ($\ln n + 2\gamma - 1$), which was proven by Peter Dirichlet (1805-59). Today, γ is called the *Euler-Mascheroni constant* (credited to both Euler and Mascheroni, $\gamma \approx 0.577215664901532860\ldots$), and appears in number theory. Perhaps the mystery of γ will be discovered in the 21st century.

October 20

S M T W T F S

There is safety in numbers
                    —anonymous

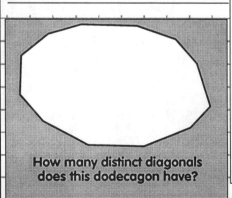

**Rearrange these ten coins
so that there are 4 coins
each in 5 straight lines.**

HISTORICAL/CURRENT NOTE:
In a guard post along the Great Wall of China a pile of wooden sticks was uncovered, which have come to be called the Han Sticks. These sticks are from the time of the Han Dynasty (ca.200BC to 200AD) and record daily activities of the Chinese living around that portion of the Great Wall. These simple wooden sticks, among other things, reveal information about troops, their duties, payments for bricks used to build the wall, and distances of operations. The numerals appearing on these sticks show tallies written from the top down, the same direction used in China today. They also reveal a shortcut-like method for writing the numerals for 20, 30, and 40; instead of the traditional way they are written .

October 21

S M T W T F S

The mathematical sciences particularly exhibit order, symmetry, and limitation; and these are the greatest forms of the beautiful.   **—Aristotle**

**How many distinct diagonals
does this dodecagon have?**

HISTORICAL/CURRENT NOTE:
It is not surprising to learn that astronomers use minutes and seconds for fractional degree measures. But it is surprising to learn this is an ancient method. When Babylonian astronomers wrote a number beginning with their zero symbol, it referred to a fractional portion of a degree, namely its minutes and seconds. Thus, the number ⪤ ⟨ Y meant 0 degrees+11/60 minutes. Greek astronomers around 200 B.C. used their alphabet numerals in a similar fashion — for example, τ ΚΑ ΒΓ meant 0+21/60 +23/360 (today we write this as 0°21'23"). Similar alphabetic notation was used by Arab and Hebrew astronomers. The sexagesimal system is still in use today.

October 22

S M T W T F S

Errors using inadequate data are much less than those using no data at all. —**Charles Babbage**

HISTORICAL/CURRENT NOTE:
Today, the 19th century *sinus curves* of John Fourier have undergone facelifts. First, the traditional sinus curve was miniaturized over equally spaced intervals, and called *wavelets*. Then, mathematician Ingrid *Daubechies'* went a step further by describing the length of wavelets over *varying* minuscule time intervals that *change according to the pitch* of the sound. These curves can describe or capture a specific sound from a conglomeration of sounds, enhance digital imagery, or develop new methods for numerical analysis. In law enforcement, wavelets lend themselves to rapid identification and transmission of fingerprint data

Arrange twelve X's on this grid so that no more than two X's appear in any row, column, or diagonal.

October 23

S M T W T F S

The study of geometry is a petty and idle exercise of the mind, if it is applied to no larger system than the starry one. —**Thoreau**

A 6 foot deer is foraging at the outskirts of a forest that is 4 miles long. The deer is startled by a hunter and starts running into the forest. How far can it run into the forest?

HISTORICAL/CURRENT NOTE:
Leonhard Euler's (1707-83) name has become synonymous with the *Königsberg bridge problem*. Yet, during his life time he published well over 500 books and papers, and hundreds of others were published after his death. His works include topics on analysis, differential and integral calculus, and the motion of celestial bodies. As would be expected, anyone writing so much would develop short hand symbols, and the mathematical symbols he introduced were i for $\sqrt{-1}$, the summation symbol Σ, and the *function symbol* $f(x)$. He is also famous for his work on polyhedra, and his formula relating a polyhedra's vertices, edges and faces, $V-E+F=2$, and the famous formula $e^{\pi}=-1$.

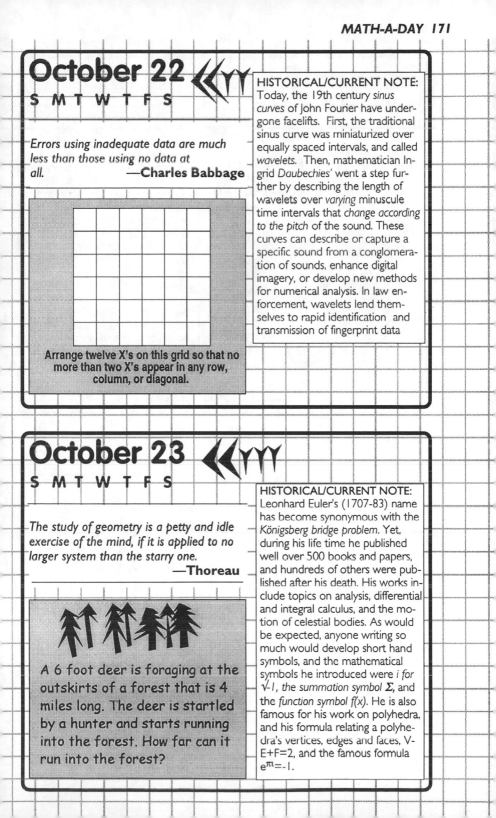

October 24

S M T W T F S

He who has heard the same thing told by 12,000 eye-witnesses has only 12,000 probabilities, which are equal to one strong probability, which is far from certain.

—Voltaire

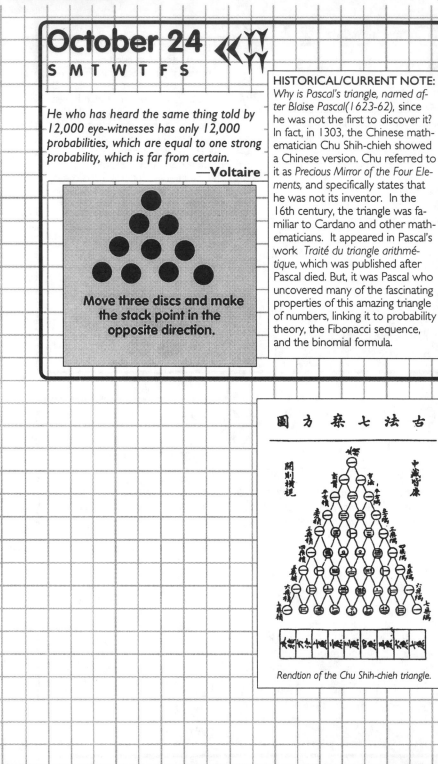

Move three discs and make the stack point in the opposite direction.

HISTORICAL/CURRENT NOTE:
Why is Pascal's triangle, named after Blaise Pascal(1623-62), since he was not the first to discover it? In fact, in 1303, the Chinese mathematician Chu Shih-chieh showed a Chinese version. Chu referred to it as *Precious Mirror of the Four Elements,* and specifically states that he was not its inventor. In the 16th century, the triangle was familiar to Cardano and other mathematicians. It appeared in Pascal's work *Traité du triangle arithmétique,* which was published after Pascal died. But, it was Pascal who uncovered many of the fascinating properties of this amazing triangle of numbers, linking it to probability theory, the Fibonacci sequence, and the binomial formula.

Rendtion of the Chu Shih-chieh triangle.

October 25

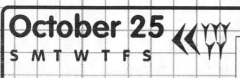

S M T W T F S

Every mathematician worthy of the name has experienced ... the state of lucid exaltation in which one thought succeeds another as if miraculously... **—Andre Weil**

A class of graduating seniors is practicing their march. If they walk in pairs, one person is without a partner. If they walk by 3s, again they are one person short. In fact, this also happens by 5s and 7s. What is the smallest number of seniors this class has?

HISTORICAL/CURRENT NOTE:
Can computers be made to think? Although the proof of the famous *four color map problem* was done by computer in 1976, it was done not by reasoning, but by the brute force of exhausting all possibilities. In 1996 a computer, for the first time, solved the famous unsolved *Robbins conjecture* by a form of reasoning! Computer scientist William McCune developed a set of programs capable of determining and deciding whether problems in abstract algebra and symbolic logic were true. The computer was also programmed to leave out steps it determined were obvious or trivial. McCune's program EQP(equation power) formulated the proof.

October 26

S M T W T F S

We are not very pleased when we are forced to accept a mathematical truth by virtue of a complicated chain of formal conclusions and computations, which we traverse blindly, link by link, feeling our way by touch. We want first an overview of the aim and of the road; we want to understand the idea of the proof, the deeper context **—Hermann Weyl**

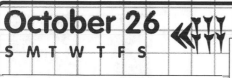

$$\left(4\tan\frac{\pi}{4} \right)\left(6\cot45^{0} \right)$$

HISTORICAL/CURRENT NOTE:
At the American Mathematical Society, in 1903, Frank Cole, one of the presenters gave a silent lecture that entailed computing manually $2^{67}-1$. After getting the result, he proceeded to multiply two very large numbers together whose product was the 67th Mersenne. After the audience had a moment to digest what they had witnessed, they gave Cole a standing ovation. He had just shown that Mersenne's 67th number was not prime, as Mersenne had contended. Marin Mersenne(1588-1648) was a mathematician and philosopher who was captivated by the quest for prime numbers. Today, the numbers generated by his formula $2^{P}-1$ are known as *Mersenne numbers*, where p is a prime number..

October 27 ◀◀ ✠✠✠
S M T W T F S

It can be shown that a mathematical web of some kind can be woven about any universe containing several objects. The fact that our universe lends itself to mathematical treatment is not a fact of any great philosophical significance.
— **Bertrand Russell**

$$(.05)\left(3^2\right)\left(\frac{20}{2^{-1}}\right)$$

HISTORICAL/CURRENT NOTE:
In the past, the study of knots was an interesting pastime. Playing with ropes, discovering magician's false knots, rigging a boat, etc.. But in the mathematics of topology , the study of knots is a serious and dynamic field. Knots, mathematically speaking, have no ends and cannot exist in more than three dimensions. Mathematicians have categorized and mathematized the properties of knots, so that today knot theory plays a very important role in physics and molecular biology. In fact, in 1999, molecular biologists Titia de Lange and Jack Griffth discovered that normal chromosomes have *no ends,* and are in fact mathematical in nature!

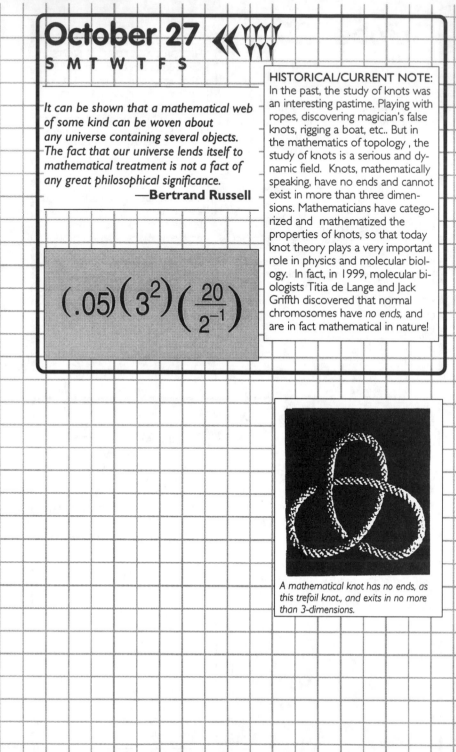

A mathematical knot has no ends, as this trefoil knot., and exits in no more than 3-dimensions.

October 28 «

S M T W T F S

Inspiration is needed in geometry, just as much as in poetry.
—Alexander Pushkin

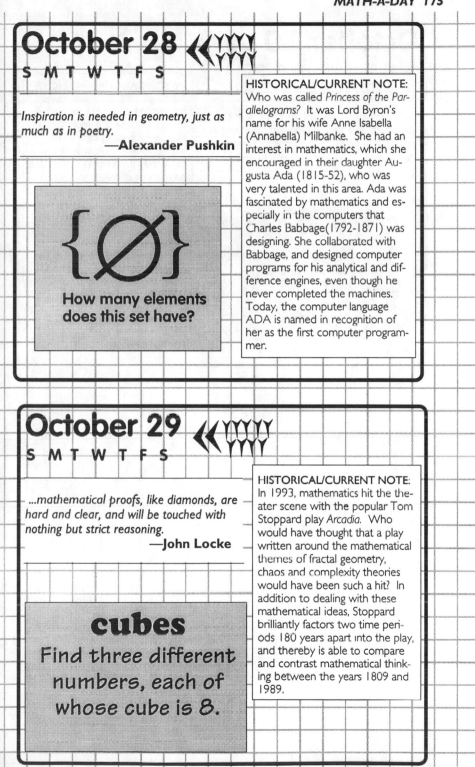

{ Ø }

How many elements does this set have?

HISTORICAL/CURRENT NOTE:
Who was called *Princess of the Parallelograms?* It was Lord Byron's name for his wife Anne Isabella (Annabella) Milbanke. She had an interest in mathematics, which she encouraged in their daughter Augusta Ada (1815-52), who was very talented in this area. Ada was fascinated by mathematics and especially in the computers that Charles Babbage(1792-1871) was designing. She collaborated with Babbage, and designed computer programs for his analytical and difference engines, even though he never completed the machines. Today, the computer language ADA is named in recognition of her as the first computer programmer.

October 29 «

S M T W T F S

...mathematical proofs, like diamonds, are hard and clear, and will be touched with nothing but strict reasoning.
—John Locke

cubes
Find three different numbers, each of whose cube is 8.

HISTORICAL/CURRENT NOTE:
In 1993, mathematics hit the theater scene with the popular Tom Stoppard play *Arcadia.* Who would have thought that a play written around the mathematical themes of fractal geometry, chaos and complexity theories would have been such a hit? In addition to dealing with these mathematical ideas, Stoppard brilliantly factors two time periods 180 years apart into the play, and thereby is able to compare and contrast mathematical thinking between the years 1809 and 1989.

October 30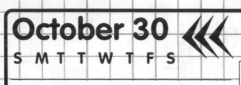

S M T T W T F S

Nothing is more important than to see the sources of invention which are, in my opinion more interesting than the inventions themselves. —**Gottfried Leibniz**

6, 18, 7, ?, 10, 30, 19, 57,...

HISTORICAL/CURRENT NOTE:
Although the soliton wave was first recorded in 1834, its incredible properties have important application in today's high tech world. Unlike most waves, soliton waves theoretically go on and on without changing shape or losing momentum, making them ideal for high speed electronic transmission of digital data along optical fibers. Equally important, soliton waves can travel near each other without affecting one another.

Diophantus used this symbol to designate subtraction.

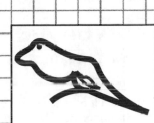

The tadpole was the Egyptian symbol for 100,000.

October 31 ◀◀◀ ▼

S M T W T F S

The further a mathematical theory is developed, the more harmoniously and uniformly does its construction proceed, and unsuspected relations are disclosed between hitherto separated branches of the science.
—**David Hilbert**

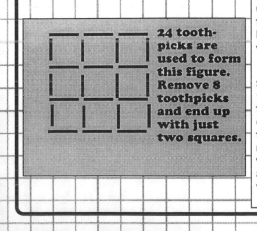

24 tooth-
picks are
used to form
this figure.
Remove 8
toothpicks
and end up
with just
two squares.

HISTORICAL/CURRENT NOTE:
Egyptian's had an interesting way of working with fractions. They used their glyph for *mouth* to indicate the numerator 1 of a fraction. Here is how they would write

Most of their fractions usually had numerators of 1 except for a few particular fractions, namely

2/3 and 3/4.

If the entire number could not fit under the *mouth*, then the rest of the number was written to the left. For example, 1/213 was written

1/4 and 1/2 had their own special symbols, namely, X = 1/4 and ⊂ = 1/2. Other than these fractions, the Egyptians would figure out how to express other fractions as the sum of fractions with numerators of 1. For example, to write 3/5, they would express it as the sum of 1/2 + 1/10. To write 7/12, they would write 1/3 + 1/4.

November 1 I

S M T W T F S

November dates are written in the Greek Attic number system

How can it be that mathematics, being after all a product of human thought independent of experience, is so admirably adapted to the objects of reality?

—Albert Einstein

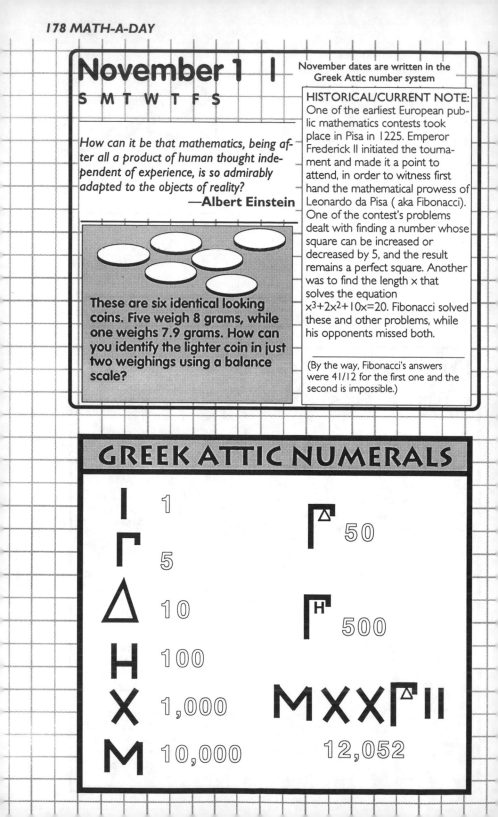

These are six identical looking coins. Five weigh 8 grams, while one weighs 7.9 grams. How can you identify the lighter coin in just two weighings using a balance scale?

HISTORICAL/CURRENT NOTE: One of the earliest European public mathematics contests took place in Pisa in 1225. Emperor Frederick II initiated the tournament and made it a point to attend, in order to witness first hand the mathematical prowess of Leonardo da Pisa (aka Fibonacci). One of the contest's problems dealt with finding a number whose square can be increased or decreased by 5, and the result remains a perfect square. Another was to find the length x that solves the equation $x^3+2x^2+10x=20$. Fibonacci solved these and other problems, while his opponents missed both.

(By the way, Fibonacci's answers were 41/12 for the first one and the second is impossible.)

GREEK ATTIC NUMERALS

Ⅰ	1	
Γ	5	
Δ	10	
H	100	
X	1,000	
M	10,000	

Γ△   50

Γ^H   500

M X X Γ△ Ⅱ

12,052

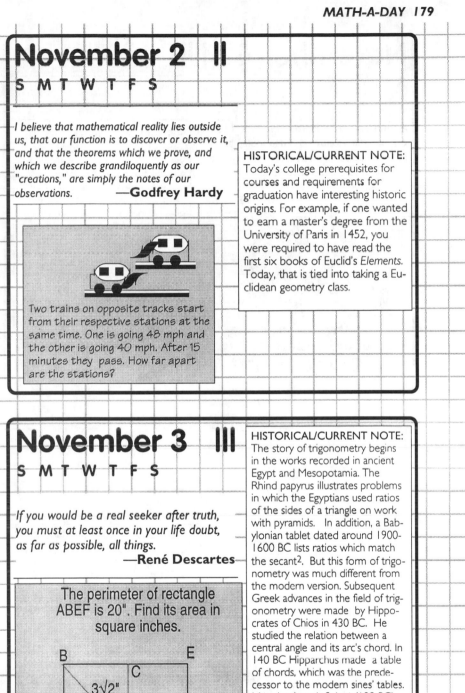

November 2 II

S M T W T F S

I believe that mathematical reality lies outside us, that our function is to discover or observe it, and that the theorems which we prove, and which we describe grandiloquently as our "creations," are simply the notes of our observations. —**Godfrey Hardy**

HISTORICAL/CURRENT NOTE: Today's college prerequisites for courses and requirements for graduation have interesting historic origins. For example, if one wanted to earn a master's degree from the University of Paris in 1452, you were required to have read the first six books of Euclid's *Elements*. Today, that is tied into taking a Euclidean geometry class.

Two trains on opposite tracks start from their respective stations at the same time. One is going 48 mph and the other is going 40 mph. After 15 minutes they pass. How far apart are the stations?

November 3 III

S M T W T F S

If you would be a real seeker after truth, you must at least once in your life doubt, as far as possible, all things. —**René Descartes**

The perimeter of rectangle ABEF is 20". Find its area in square inches.

B   E

$3\sqrt{2}$"

C

F

A   D

HISTORICAL/CURRENT NOTE: The story of trigonometry begins in the works recorded in ancient Egypt and Mesopotamia. The Rhind papyrus illustrates problems in which the Egyptians used ratios of the sides of a triangle on work with pyramids. In addition, a Babylonian tablet dated around 1900-1600 BC lists ratios which match the secant[2]. But this form of trigonometry was much different from the modern version. Subsequent Greek advances in the field of trigonometry were made by Hippocrates of Chios in 430 BC. He studied the relation between a central angle and its arc's chord. In 140 BC Hipparchus made a table of chords, which was the predecessor to the modern sines' tables. Menelaus' work *Spheric*(100 BC) introduced spherical trig. and Ptolemy's book *Almagest* (140 AD) even investigated trigonometric identities.

November 4   IIII
S M T W T F S

A man should be learned in several sciences, and should have a reasonable, philosophical and in some measure a mathematical head, to be a complete and excellent poet.

—John Dryden

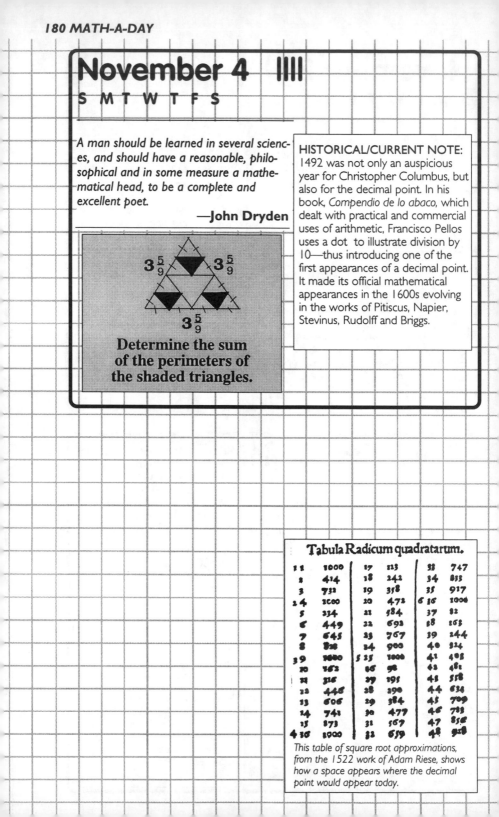

$3\frac{5}{9}$ $3\frac{5}{9}$

$3\frac{5}{9}$

Determine the sum of the perimeters of the shaded triangles.

HISTORICAL/CURRENT NOTE:
1492 was not only an auspicious year for Christopher Columbus, but also for the decimal point. In his book, *Compendio de lo abaco*, which dealt with practical and commercial uses of arithmetic, Francisco Pellos uses a dot to illustrate division by 10—thus introducing one of the first appearances of a decimal point. It made its official mathematical appearances in the 1600s evolving in the works of Pitiscus, Napier, Stevinus, Rudolff and Briggs.

Tabula Radicum quadratarum.

1 1	1000	17	123	33	747
2	414	18	242	34	833
3	732	19	358	35	917
2 4	2000	20	472	6 36	1000
5	234	21	584	37	82
6	449	22	692	38	163
7	645	23	767	39	244
8	828	24	900	40	324
3 9	1000	5 25	1000	41	405
10	162	26	98	42	481
11	316	27	195	43	558
12	448	28	290	44	634
13	606	29	384	45	709
14	741	30	477	46	783
15	873	31	567	47	856
4 16	1000	32	659	48	928

This table of square root approximations, from the 1522 work of Adam Riese, shows how a space appears where the decimal point would appear today.

November 5 Γ

S M T W T F S

The mathematician has reached the highest rung on the ladder of human thought. —**Havelock Ellis**

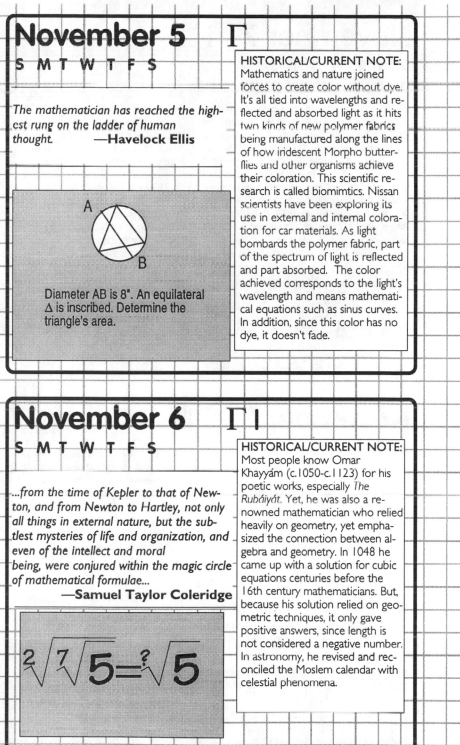

A

B

Diameter AB is 8". An equilateral Δ is inscribed. Determine the triangle's area.

HISTORICAL/CURRENT NOTE: Mathematics and nature joined forces to create color without dye. It's all tied into wavelengths and reflected and absorbed light as it hits two kinds of new polymer fabrics being manufactured along the lines of how iridescent Morpho butterflies and other organisms achieve their coloration. This scientific research is called biomimtics. Nissan scientists have been exploring its use in external and internal coloration for car materials. As light bombards the polymer fabric, part of the spectrum of light is reflected and part absorbed. The color achieved corresponds to the light's wavelength and means mathematical equations such as sinus curves. In addition, since this color has no dye, it doesn't fade.

November 6 Γι

S M T W T F S

...from the time of Kepler to that of Newton, and from Newton to Hartley, not only all things in external nature, but the subtlest mysteries of life and organization, and even of the intellect and moral being, were conjured within the magic circle of mathematical formulae... —**Samuel Taylor Coleridge**

$$\sqrt[2]{\sqrt[7]{5}} = \sqrt[?]{5}$$

HISTORICAL/CURRENT NOTE: Most people know Omar Khayyám (c.1050-c.1123) for his poetic works, especially *The Rubáiyát*. Yet, he was also a renowned mathematician who relied heavily on geometry, yet emphasized the connection between algebra and geometry. In 1048 he came up with a solution for cubic equations centuries before the 16th century mathematicians. But, because his solution relied on geometric techniques, it only gave positive answers, since length is not considered a negative number. In astronomy, he revised and reconciled the Moslem calendar with celestial phenomena.

November 7 Γ ‖

S M T W T F S

Philosophy is written in this grand book—I mean the universe—which stands continually open to our gaze, but it cannot be understood unless one first learns to comprehend the language and interpret the characters in which it is written. It is written in the language of mathematics, and its characters are triangles, circles, and other geometric figures, without which it is humanly impossible to understand a single word of it; without these, one is wandering in a dark labyrinth. .

— **Galileo Galilei**

$$\left(\left(\frac{\sqrt[3]{3}}{64}\right)(8)^2\right)^3$$

HISTORICAL/CURRENT NOTE: History has proven time and again, that some mathematical problem's solution is *no solution*— e.g. the *Königsberg bridge problem*, the *three ancient construction problems.* In 1824 Niels Abel (1802-1829), in his quest to find a general solution for 5th degree equations along the same lines as those proven for the quadratic and cubic equations, proved no such algebraic solution (involving radicals) was possible. He was only 19 at the time. In 1799, a rather vague, less refined and little known proof had been done by Paolo Ruffini (1765-1822).

November 8 Γ ‖‖

S M T W T F S

Since you are now studying geometry and trigonometry, I will give you a problem. A ship sails the ocean. It left Boston with a cargo of wool. It grosses 200 tons. It is bound for Le Havre. The main mast is broken, the cabin boy is on deck, there are 12 passengers aboard, the wind is blowing East-North-East, the clock points to a quarter past three in the afternoon. It is the month of May. How old is the captain?

— **Gustave Flaubert**

$$X - 202 = 0$$
seventeen   four

HISTORICAL/CURRENT NOTE: Abel was an exceptionally talented mathematician, but like Galois and Ramanujan, an early death cut his career short. In 1826 Abel went to Paris to seek a mathematics professorship. He left a manuscript with Augustin Cauchy for the French Academy of Sciences, which he hoped would demonstrate his abilities and originality. Unfortunately, Cauchy put the manuscript aside, and, in fact, misplaced it. Abel was fortunate to have a German journal, *Crelle's journal*, print five of his papers. In 1826, two days after Abel died of tuberculosis, he received a letter offering him a professorship in mathematics at the University of Berlin.

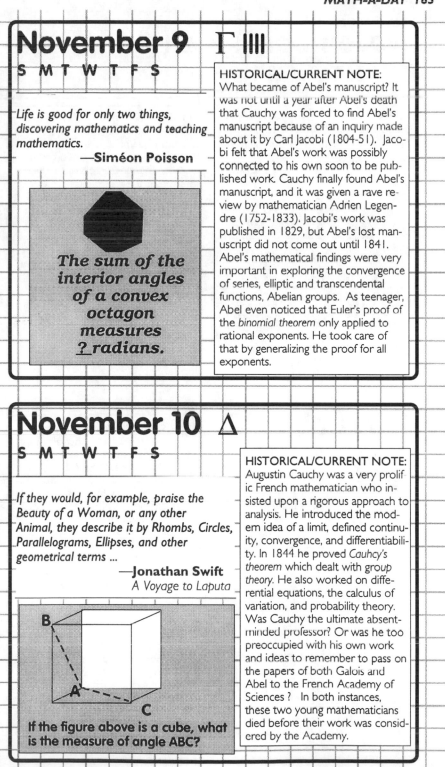

November 9 Γ ||||
S M T W T F S

*Life is good for only two things,
discovering mathematics and teaching
mathematics.*
—**Siméon Poisson**

**The sum of the
interior angles
of a convex
octagon
measures
? radians.**

HISTORICAL/CURRENT NOTE:
What became of Abel's manuscript? It
was not until a year after Abel's death
that Cauchy was forced to find Abel's
manuscript because of an inquiry made
about it by Carl Jacobi (1804-51). Jacobi felt that Abel's work was possibly
connected to his own soon to be published work. Cauchy finally found Abel's
manuscript, and it was given a rave review by mathematician Adrien Legendre (1752-1833). Jacobi's work was
published in 1829, but Abel's lost manuscript did not come out until 1841.
Abel's mathematical findings were very
important in exploring the convergence
of series, elliptic and transcendental
functions, Abelian groups. As teenager,
Abel even noticed that Euler's proof of
the *binomial theorem* only applied to
rational exponents. He took care of
that by generalizing the proof for all
exponents.

November 10 Δ
S M T W T F S

*If they would, for example, praise the
Beauty of a Woman, or any other
Animal, they describe it by Rhombs, Circles,
Parallelograms, Ellipses, and other
geometrical terms ...*
—**Jonathan Swift**
A Voyage to Laputa

**If the figure above is a cube, what
is the measure of angle ABC?**

HISTORICAL/CURRENT NOTE:
Augustin Cauchy was a very prolific French mathematician who insisted upon a rigorous approach to
analysis. He introduced the modem idea of a limit, defined continuity, convergence, and differentiabili-
ty. In 1844 he proved *Cauhcy's
theorem* which dealt with group
theory. He also worked on differential equations, the calculus of
variation, and probability theory.
Was Cauchy the ultimate absentminded professor? Or was he too
preoccupied with his own work
and ideas to remember to pass on
the papers of both Galois and
Abel to the French Academy of
Sciences ? In both instances,
these two young mathematicians
died before their work was considered by the Academy.

November 11 △ I

S M T W T F S

...there is no study in the world which brings into more harmonious action all the faculties of the mind than [mathematics], ...

— **J.J. Sylvester**

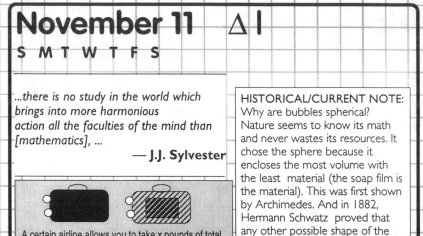

A certain airline allows you to take x pounds of total baggage free of charge. For each pound over x you are changed a fix amount per pound. Suppose Mary and Karen each packed over this amount. Their bags all together weighed 70 pounds. For her extra bag weight, Mary had to pay $3.50, and Karen paid $1.50. If just one person checked the weight of their bags, that person would have had to pay $20 for the extra weight. How many pounds of baggage did Mary and Karen each bring? What was the charge for each extra pound? What was the allowed number of pounds?

HISTORICAL/CURRENT NOTE:
Why are bubbles spherical? Nature seems to know its math and never wastes its resources. It chose the sphere because it encloses the most volume with the least material (the soap film is the material). This was first shown by Archimedes. And in 1882, Hermann Schwatz proved that any other possible shape of the imagination didn't work as well as the sphere.

November 12 △II

S M T T W T F S

Can we actually "know" the universe? My God, it's hard enough finding your way around Chinatown.

—**Woody Allen**

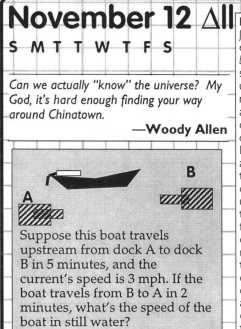

Suppose this boat travels upstream from dock A to dock B in 5 minutes, and the current's speed is 3 mph. If the boat travels from B to A in 2 minutes, what's the speed of the boat in still water?

HISTORICAL/CURRENT NOTE:
Joel Hass and Roger Schlafly wanted to make sure that nature's *double bubble* was the most efficient shape for enclosing two equal volumes. In 1995, they put mathematics and a computer to work and showed there were only two contenders for the answer—the *double bubble* and the *torus bubble*. In the process, they came up with a way to bypass computer errors that might accumulate by rounding off decimal answers of the measured volumes. They programmed the computer to compare results from both shapes and keep tabs of rounded off answers. The naturally occurring *double bubble*, as nature already knew, was the most efficient shape. The mathematically created *torus bubble* lost out. In addition, Schlafly and Hass' method is a new method for finding the maximum or minimum of various things.

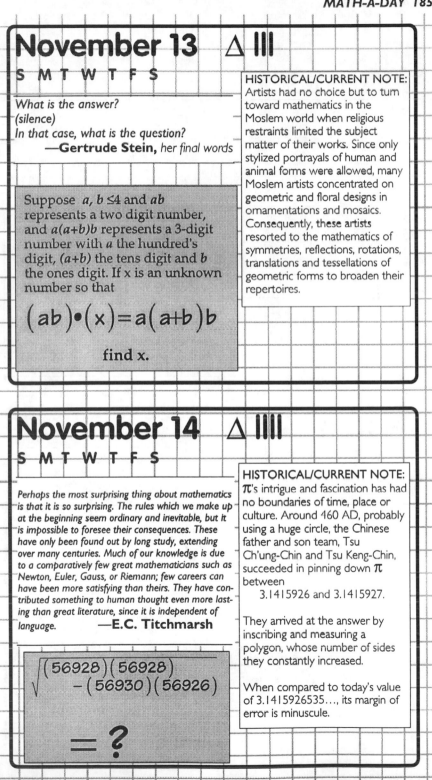

November 13 △ |||

S M T W T F S

What is the answer?
(silence)
In that case, what is the question?
 —Gertrude Stein, *her final words*

Suppose *a, b* ≤4 and *ab* represents a two digit number, and *a(a+b)b* represents a 3-digit number with *a* the hundred's digit, *(a+b)* the tens digit and *b* the ones digit. If x is an unknown number so that

$$\left(ab\right)\bullet\left(x\right)=a\left(a+b\right)b$$

find x.

HISTORICAL/CURRENT NOTE:
Artists had no choice but to turn toward mathematics in the Moslem world when religious restraints limited the subject matter of their works. Since only stylized portrayals of human and animal forms were allowed, many Moslem artists concentrated on geometric and floral designs in ornamentations and mosaics. Consequently, these artists resorted to the mathematics of symmetries, reflections, rotations, translations and tessellations of geometric forms to broaden their repertoires.

November 14 △ ||||

S M T W T F S

Perhaps the most surprising thing about mathematics is that it is so surprising. The rules which we make up at the beginning seem ordinary and inevitable, but it is impossible to foresee their consequences. These have only been found out by long study, extending over many centuries. Much of our knowledge is due to a comparatively few great mathematicians such as Newton, Euler, Gauss, or Riemann; few careers can have been more satisfying than theirs. They have contributed something to human thought even more lasting than great literature, since it is independent of language. **—E.C. Titchmarsh**

$$\sqrt{\begin{array}{c}\left(56928\right)\left(56928\right)\\ -\left(56930\right)\left(56926\right)\end{array}}$$

$$= \text{?}$$

HISTORICAL/CURRENT NOTE:
π's intrigue and fascination has had no boundaries of time, place or culture. Around 460 AD, probably using a huge circle, the Chinese father and son team, Tsu Ch'ung-Chin and Tsu Keng-Chin, succeeded in pinning down π between
 3.1415926 and 3.1415927.

They arrived at the answer by inscribing and measuring a polygon, whose number of sides they constantly increased.

When compared to today's value of 3.1415926535…, its margin of error is minuscule.

Rendition of two of the many moon phases as
they appeared to Galileo through his telescope.

November 15 $\Delta\Gamma$

S M T W T F S

*Mathematical discoveries, small or great
are never born of spontaneous generation.
They always presuppose a soil seeded with
preliminary knowledge and well prepared
by labour, both conscious and
subconscious.*

—Henri Poincaré

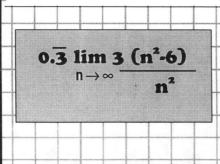

$$0.\overline{3} \lim_{n \to \infty} \frac{3\,(n^2\text{-}6)}{n^2}$$

HISTORICAL/CURRENT NOTE:
When Galileo's name is mentioned—the tower of Pisa—
astronomical discoveries—the
Inquisition— immediately come to
mind. Born in Pisa in 1564, he
began his career there as a
Professor of Mathematics. He
viewed mathematics as the language, not only of the sciences, but
of the universe. His astronomical
observations, specifically the
moons orbiting Jupiter, gave physical credence to Copernicus'
theory of the universe, which had
the planets orbiting the Sun rather
than having Earth as the center of
all. For his discoveries he was hon-
ored with an appointment as Chief
Mathematician and Philosopher to
Cosimo de Medici in Florence in
1610.

November 16 ΔΓΙ

S M T W T F S

Science has explored the microcosmos and the macrocosmos...The great unexplored frontier is complexity...I am convinced that nations and people that master the new science of Complexity will become the economic, cultural, and political superpowers of the next century.

—**Heinz Pagels**

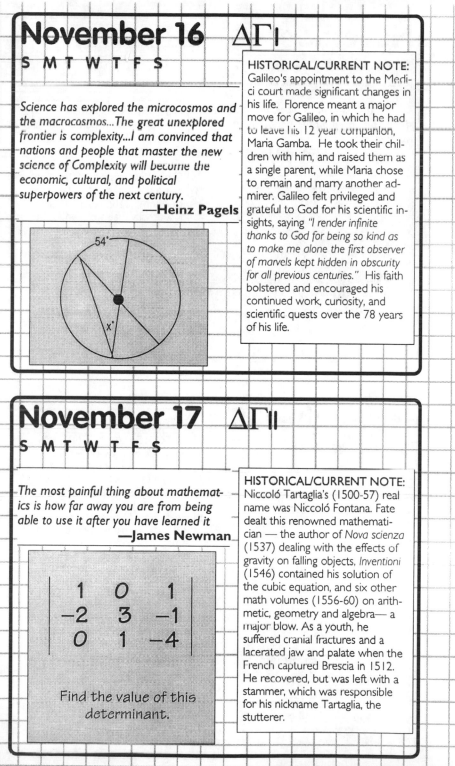

54°

x°

HISTORICAL/CURRENT NOTE: Galileo's appointment to the Medici court made significant changes in his life. Florence meant a major move for Galileo, in which he had to leave his 12 year companion, Maria Gamba. He took their children with him, and raised them as a single parent, while Maria chose to remain and marry another admirer. Galileo felt privileged and grateful to God for his scientific insights, saying *"I render infinite thanks to God for being so kind as to make me alone the first observer of marvels kept hidden in obscurity for all previous centuries."* His faith bolstered and encouraged his continued work, curiosity, and scientific quests over the 78 years of his life.

November 17 ΔΓΙΙ

S M T W T F S

The most painful thing about mathematics is how far away you are from being able to use it after you have learned it

—**James Newman**

$$\begin{vmatrix} 1 & 0 & 1 \\ -2 & 3 & -1 \\ 0 & 1 & -4 \end{vmatrix}$$

Find the value of this determinant.

HISTORICAL/CURRENT NOTE: Niccoló Tartaglia's (1500-57) real name was Niccoló Fontana. Fate dealt this renowned mathematician — the author of *Nova scienza* (1537) dealing with the effects of gravity on falling objects, *Inventioni* (1546) contained his solution of the cubic equation, and six other math volumes (1556-60) on arithmetic, geometry and algebra— a major blow. As a youth, he suffered cranial fractures and a lacerated jaw and palate when the French captured Brescia in 1512. He recovered, but was left with a stammer, which was responsible for his nickname Tartaglia, the stutterer.

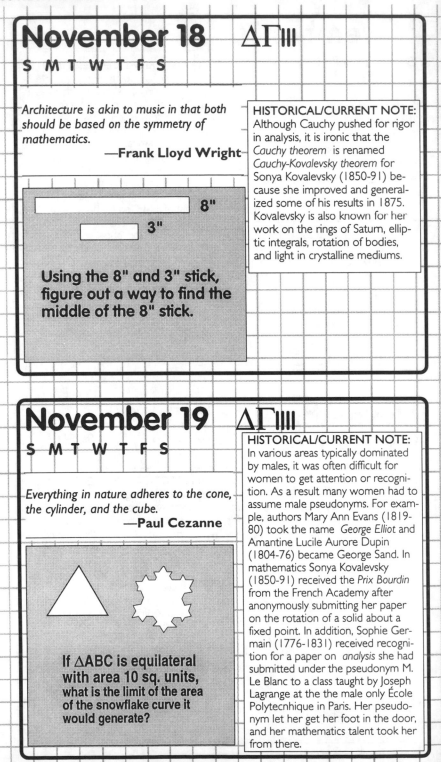

November 18 ΔΓ||||

S M T W T F S

Architecture is akin to music in that both should be based on the symmetry of mathematics.

—Frank Lloyd Wright

8"

3"

Using the 8" and 3" stick, figure out a way to find the middle of the 8" stick.

HISTORICAL/CURRENT NOTE:
Although Cauchy pushed for rigor in analysis, it is ironic that the *Cauchy theorem* is renamed *Cauchy-Kovalevsky theorem* for Sonya Kovalevsky (1850-91) because she improved and generalized some of his results in 1875. Kovalevsky is also known for her work on the rings of Saturn, elliptic integrals, rotation of bodies, and light in crystalline mediums.

November 19 ΔΓ||||

S M T W T F S

Everything in nature adheres to the cone, the cylinder, and the cube.

—Paul Cezanne

If △ABC is equilateral with area 10 sq. units, what is the limit of the area of the snowflake curve it would generate?

HISTORICAL/CURRENT NOTE:
In various areas typically dominated by males, it was often difficult for women to get attention or recognition. As a result many women had to assume male pseudonyms. For example, authors Mary Ann Evans (1819-80) took the name *George Elliot* and Amantine Lucile Aurore Dupin (1804-76) became George Sand. In mathematics Sonya Kovalevsky (1850-91) received the *Prix Bourdin* from the French Academy after anonymously submitting her paper on the rotation of a solid about a fixed point. In addition, Sophie Germain (1776-1831) received recognition for a paper on *analysis* she had submitted under the pseudonym M. Le Blanc to a class taught by Joseph Lagrange at the the male only École Polytecnique in Paris. Her pseudonym let her get her foot in the door, and her mathematics talent took her from there.

November 20 △△

S M T W T F S

Joy and amazement at the beauty and grandeur of this world of which man can just form a faint notion.

—**Albert Einstein**

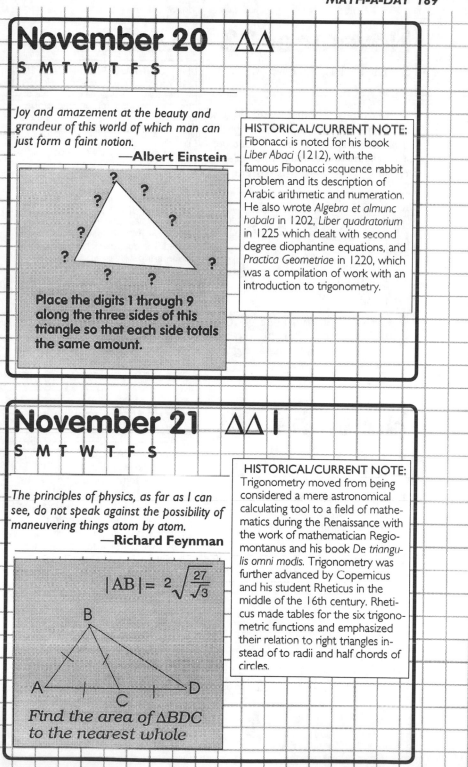

Place the digits 1 through 9 along the three sides of this triangle so that each side totals the same amount.

HISTORICAL/CURRENT NOTE: Fibonacci is noted for his book *Liber Abaci* (1212), with the famous Fibonacci sequence rabbit problem and its description of Arabic arithmetic and numeration. He also wrote *Algebra et almunc habala* in 1202, *Liber quadratorium* in 1225 which dealt with second degree diophantine equations, and *Practica Geometriae* in 1220, which was a compilation of work with an introduction to trigonometry.

November 21 △△I

S M T W T F S

The principles of physics, as far as I can see, do not speak against the possibility of maneuvering things atom by atom.

—**Richard Feynman**

$$|AB| = 2\sqrt{\frac{27}{\sqrt{3}}}$$

Find the area of △BDC to the nearest whole

HISTORICAL/CURRENT NOTE: Trigonometry moved from being considered a mere astronomical calculating tool to a field of mathematics during the Renaissance with the work of mathematician Regiomontanus and his book *De triangulis omni modis*. Trigonometry was further advanced by Copernicus and his student Rheticus in the middle of the 16th century. Rheticus made tables for the six trigonometric functions and emphasized their relation to right triangles instead of to radii and half chords of circles.

November 22 △△ ‖

S M T W T F S

One must regard nature reasonably and naturally as one would truth, and be contented only with a representation of it which errs to the smallest possible extent. —**Janos Bolyai**

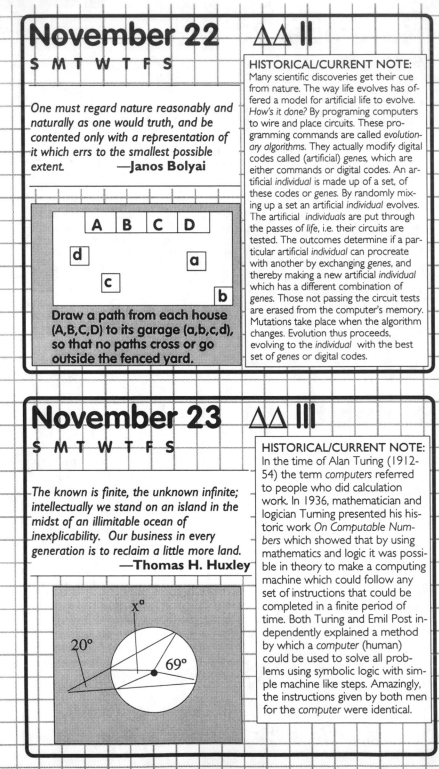

| A | B | C | D |

d

a

c

b

Draw a path from each house (A,B,C,D) to its garage (a,b,c,d), so that no paths cross or go outside the fenced yard.

HISTORICAL/CURRENT NOTE:
Many scientific discoveries get their cue from nature. The way life evolves has offered a model for artificial life to evolve. *How's it done?* By programing computers to wire and place circuits. These programming commands are called *evolutionary algorithms.* They actually modify digital codes called (artificial) *genes,* which are either commands or digital codes. An artificial *individual* is made up of a set, of these codes or *genes.* By randomly mixing up a set an artificial *individual* evolves. The artificial *individuals* are put through the passes of *life,* i.e. their circuits are tested. The outcomes determine if a particular artificial *individual* can procreate with another by exchanging *genes,* and thereby making a new artificial *individual* which has a different combination of *genes.* Those not passing the circuit tests are erased from the computer's memory. Mutations take place when the algorithm changes. Evolution thus proceeds, evolving to the *individual* with the best set of *genes* or digital codes.

November 23 △△ ‖‖

S M T W T F S

The known is finite, the unknown infinite; intellectually we stand on an island in the midst of an illimitable ocean of inexplicability. Our business in every generation is to reclaim a little more land. —**Thomas H. Huxley**

20°

x°

69°

HISTORICAL/CURRENT NOTE:
In the time of Alan Turing (1912-54) the term *computers* referred to people who did calculation work. In 1936, mathematician and logician Turing presented his historic work *On Computable Numbers* which showed that by using mathematics and logic it was possible in theory to make a computing machine which could follow any set of instructions that could be completed in a finite period of time. Both Turing and Emil Post independently explained a method by which a *computer* (human) could be used to solve all problems using symbolic logic with simple machine like steps. Amazingly, the instructions given by both men for the *computer* were identical.

November 24 △△ ||||
S M T W T F S

Let chaos storm!
Let cloud shapes swarm!
I wait for form.
— **Robert Frost**

HISTORICAL/CURRENT NOTE:
George Riemann (1826-66) had a knack for introducing startling mathematical ideas in lectures. In his famous 1854 lecture, *On The Hypotheses That Lie At The Foundation Of Geometry,* he brought to light his elliptic non-Euclidean geometry. In a subsequent lecture, he gave an example of a continuous function that had no tangent lines (i.e. derivatives). Traditional mathematicians refused to take such functions seriously, and referred to them as "pathological" or "monstrous".

The tracks below are 1 yard apart. If one trolley runs along track A and another along B,

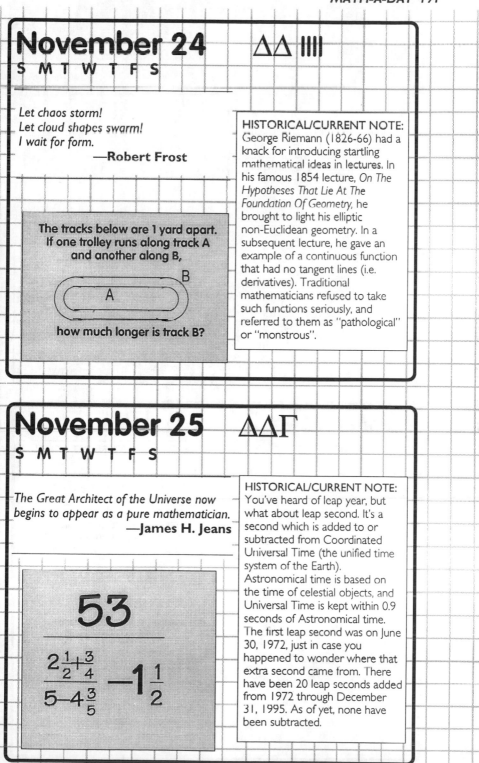

how much longer is track B?

November 25 △△Γ
S M T W T F S

The Great Architect of the Universe now begins to appear as a pure mathematician.
— **James H. Jeans**

HISTORICAL/CURRENT NOTE:
You've heard of leap year, but what about leap second. It's a second which is added to or subtracted from Coordinated Universal Time (the unified time system of the Earth). Astronomical time is based on the time of celestial objects, and Universal Time is kept within 0.9 seconds of Astronomical time. The first leap second was on June 30, 1972, just in case you happened to wonder where that extra second came from. There have been 20 leap seconds added from 1972 through December 31, 1995. As of yet, none have been subtracted.

$$\frac{53}{\dfrac{2\frac{1}{2}+\frac{3}{4}}{5-4\frac{3}{5}}} \quad -1\frac{1}{2}$$

A portion of the 1872 fractal created by Karl Weierstrass.

An enlarged section of it.

November 26 ΔΔΓΙ

S M T W T F S

Everyone knows what a curve is, until he has studied enough mathematics to become confused through the countless number of possible exceptions.

—Felix Klein

$$\sqrt[3]{8^{x-3}}=16$$

HISTORICAL/CURRENT NOTE:
"i before e except after c" It often seems that there are exceptions to rules, be they rules for grammar or mathematics. For example, *Mathematical exceptions* were what conventional mathematicians called the curves or functions (now known as fractals) which such mathematicians as Wierstrass, Cantor, and Koch studied in the 19th century. Until these unusual curves appeared, the rules were— •*all continuous curves (those without any gaps in them) always had tangents lines.* •*there were no shapes whose area was finite but perimeter infinite,* •*there were no 1-dimensional objects that could eventually cover a 2-dimensional object.* But fractals created exceptions to these well established mathematical rules. Today, these are no longer exceptions, but are part of *fractal geometry.*

November 27   ΔΔΓII
S M T W T F S

Mathematics takes us into the region of absolute necessity, to which not only the actual world, but every possible world, must conform.

—Bertrand Russell

$$y = 3\cos 4\theta - 1$$

Graph this sinus curve.

November 28   ΔΔΓIII
S M T W T F S

One need not know the profoundest mysteries of geometry to be able to discern its usefulness.

—Robert Boyle

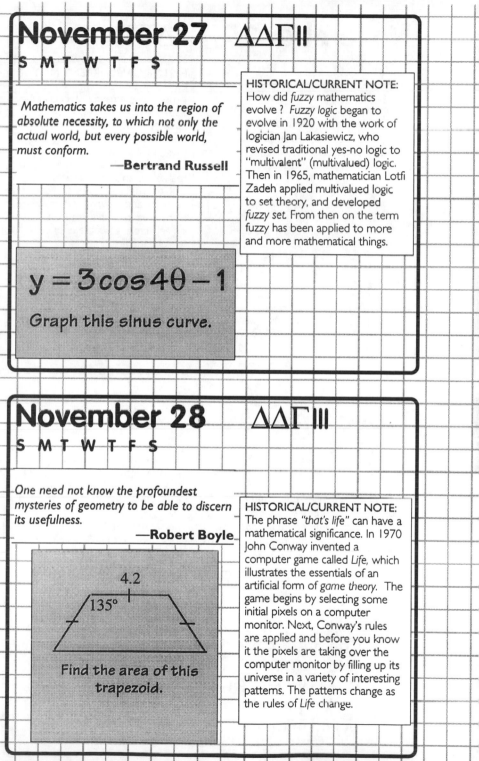

4.2

135°

Find the area of this trapezoid.

November 29

S M T W T F S

ΔΔΓ ||||

I recommend that you question all your beliefs except that two and two make four. —**Voltaire**

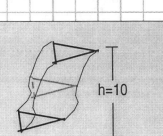

h=10

If every cross section of this prism is an isosceles right Δ with hypotenuse 6, find the prism's volume.

HISTORICAL/CURRENT NOTE:
One of the most complex and famous mathematical objects in the 20th century is the *Mandelbrot set*. Its mysterious shape has been popularized on T-shirts, magnets, mugs, posters, coasters. The more closely you look at it, the more you realize it is an infinite universe within an infinite universe, ever changing and growing. Who exactly is the creator of this beautiful object remains a battle of priority. It is not the first time mathematicians have fought over objects and ideas. Among the famous from the past are the cycloid, sometimes referred to as *the Helen of geometry*, the solution to the cubic equation, and the idea of calculus.

November 30

S M T W T F S

ΔΔΔ

The description of right lines and circles, upon which geometry is founded, belongs to mechanics. Geometry does not teach us to draw these lines, but requires them to be drawn. —**Isaac Newton**

The area of the rotating circle is (625/64)π sq.m. Find the length of this cycloid in meters.

HISTORICAL/CURRENT NOTE:
The phrase *"getting connected"* has gone global. In September of 1999 Microsoft and Softbank (a Japanese Software maker) announced a joint venture to build a fiber-optic Asian network which will go over land and into the ocean connecting Japan, China, Singapore, Hong Kong, Taiwan, South Korea, Malaysia and the Philippines.

December 1

S M T W T F S

God geometrizes.

—**Plato**

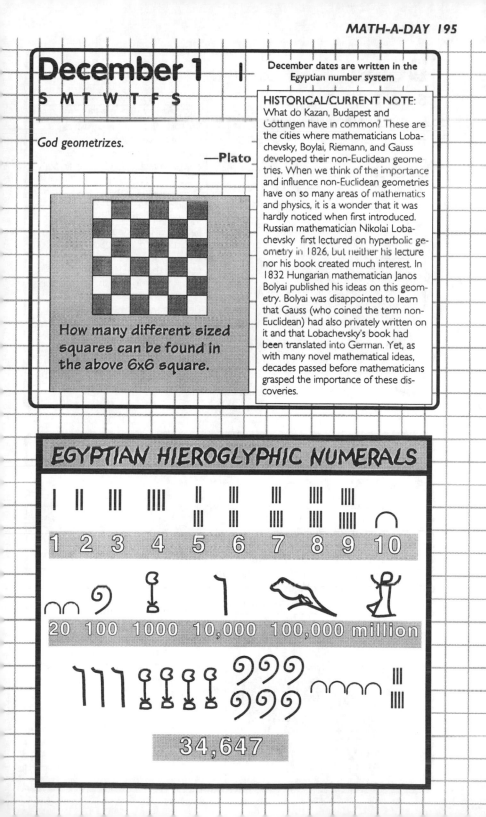

How many different sized squares can be found in the above 6x6 square.

December dates are written in the Egyptian number system

HISTORICAL/CURRENT NOTE:
What do Kazan, Budapest and Göttingen have in common? These are the cities where mathematicians Lobachevsky, Boylai, Riemann, and Gauss developed their non-Euclidean geometries. When we think of the importance and influence non-Euclidean geometries have on so many areas of mathematics and physics, it is a wonder that it was hardly noticed when first introduced. Russian mathematician Nikolai Lobachevsky first lectured on hyperbolic geometry in 1826, but neither his lecture nor his book created much interest. In 1832 Hungarian mathematician Janos Bolyai published his ideas on this geometry. Bolyai was disappointed to learn that Gauss (who coined the term non-Euclidean) had also privately written on it and that Lobachevsky's book had been translated into German. Yet, as with many novel mathematical ideas, decades passed before mathematicians grasped the importance of these discoveries.

EGYPTIAN HIEROGLYPHIC NUMERALS

1 2 3 4 5 6 7 8 9 10

20 100 1000 10,000 100,000 million

34,647

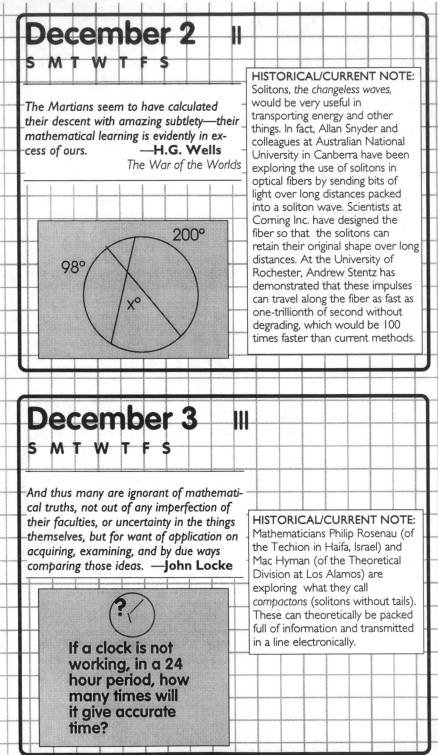

December 2 II

S M T W T F S

The Martians seem to have calculated their descent with amazing subtlety—their mathematical learning is evidently in ex-cess of ours.　　　　**—H.G. Wells**
The War of the Worlds

200°

98°

x°

HISTORICAL/CURRENT NOTE:
Solitons, *the changeless waves,* would be very useful in transporting energy and other things. In fact, Allan Snyder and colleagues at Australian National University in Canberra have been exploring the use of solitons in optical fibers by sending bits of light over long distances packed into a soliton wave. Scientists at Corning Inc. have designed the fiber so that the solitons can retain their original shape over long distances. At the University of Rochester, Andrew Stentz has demonstrated that these impulses can travel along the fiber as fast as one-trillionth of second without degrading, which would be 100 times faster than current methods.

December 3 III

S M T W T F S

And thus many are ignorant of mathemati-cal truths, not out of any imperfection of their faculties, or uncertainty in the things themselves, but for want of application on acquiring, examining, and by due ways comparing those ideas. **—John Locke**

?

If a clock is not working, in a 24 hour period, how many times will it give accurate time?

HISTORICAL/CURRENT NOTE:
Mathematicians Philip Rosenau (of the Techion in Haifa, Israel) and Mac Hyman (of the Theoretical Division at Los Alamos) are exploring what they call *compactons* (solitons without tails). These can theoretically be packed full of information and transmitted in a line electronically.

December 4

S M T W T F S

IIII

The sciences, even the best—mathematics and astronomy— are like sportsmen, who seize whatever prey offers, even without being able to make any use of it
—Ralph Waldo Emerson

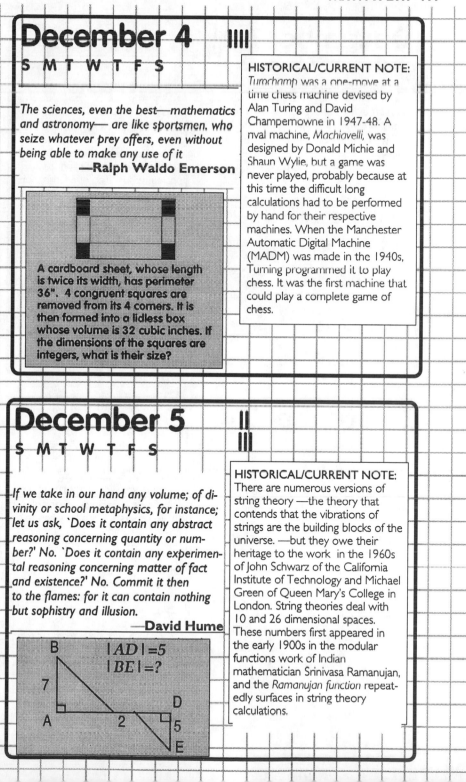

A cardboard sheet, whose length is twice its width, has perimeter 36". 4 congruent squares are removed from its 4 corners. It is then formed into a lidless box whose volume is 32 cubic inches. If the dimensions of the squares are integers, what is their size?

HISTORICAL/CURRENT NOTE:
Turochamp was a one-move at a time chess machine devised by Alan Turing and David Champernowne in 1947-48. A rival machine, *Machiavelli*, was designed by Donald Michie and Shaun Wylie, but a game was never played, probably because at this time the difficult long calculations had to be performed by hand for their respective machines. When the Manchester Automatic Digital Machine (MADM) was made in the 1940s, Turing programmed it to play chess. It was the first machine that could play a complete game of chess.

December 5

S M T W T F S

II
III

If we take in our hand any volume; of divinity or school metaphysics, for instance; let us ask, `Does it contain any abstract reasoning concerning quantity or number?' No. `Does it contain any experimental reasoning concerning matter of fact and existence?' No. Commit it then to the flames: for it can contain nothing but sophistry and illusion.
—David Hume

$|AD|=5$
$|BE|=?$

B
7
A   2   D
5
E

HISTORICAL/CURRENT NOTE:
There are numerous versions of string theory —the theory that contends that the vibrations of strings are the building blocks of the universe. —but they owe their heritage to the work in the 1960s of John Schwarz of the California Institute of Technology and Michael Green of Queen Mary's College in London. String theories deal with 10 and 26 dimensional spaces. These numbers first appeared in the early 1900s in the modular functions work of Indian mathematician Srinivasa Ramanujan, and the *Ramanujan function* repeatedly surfaces in string theory calculations.

December 6
S M T W T F S ⦀⦀

Intuition is the conception of an attentive mind, so clear, so distinct, and so effortless that we cannot doubt what we have so conceived.

—René Descartes

HISTORICAL/CURRENT NOTE: String theory deals with some very unusual concepts, such as 10 and 26 dimensional space and *compacted worlds*. To get a feeling for what a *compacted world* is, imagine a world of zero-dimension. Suppose the 0-D world actually resides in a 3-D world that is compressed into what appears to be a single point. The 3-D sphere's radius is so small, it cannot be measured, so it appears to be zero-dimensional.

$$\frac{2\sqrt{.81}}{\sqrt{.01}}$$

December 7
S M T W T F S ⦀⦀

They supposed the elements of numbers to be the elements of all things, and the whole heaven to be a musical scale and a number. **—Aristotle**
about the Pythagoreans

HISTORICAL/CURRENT NOTE: How did the ancients square-off their corners? Since they had to rely on the mathematics of their time, they were left with essentially three ways for making a right angle — (1) make a perpendicular bisector of a line segment using only a straightedge and compass; (2) use knotted ropes to form triangles with sides 3, 4, 5; (3) make a semicircle from a circle and then choose any point on the arc of the semicircle as a vertex of a right triangle whose base is the circle's diameter

$$\left(\frac{-1+i\sqrt{3}}{2}\right)^3$$

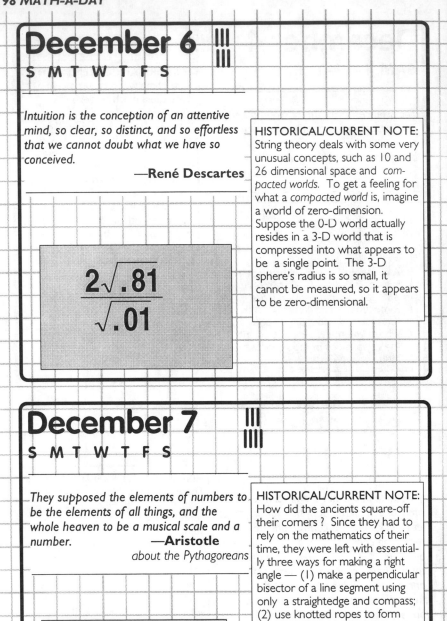

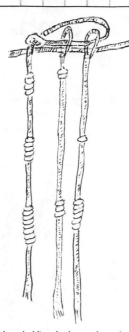

The loop holding the knotted numbers 458, 203 and 16 signifies their sum on the quipu.

December 8

S M T W T F S

A good teacher should understand and impress on his students the view that no problem whatever is completely exhausted. ... not to give his students the impression that mathematical problems have little connection with each other, and no connection at all with anything else. —**George Polyá**

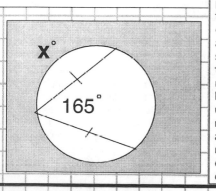

HISTORICAL/CURRENT NOTE:
On their quipus the Incas used a special method of knotting to signify various amounts . I was signified by a single knot, 2 a figure eight knot, 3 had three loops on a type of slip knot; 2 to 9 had two to nine loops on a slip knot; the first knot near the bottom of the cord was the ones place, the next row up was the tens, the next above that was the hundreds, a space was zero. So if 458 appears in the first cord; 203 in the second cord; 16 is on the third cord; the cord holding these three cords represents the sum of these numbers, namely 677. The scribes also used knots of quipus to record non-numerical information somewhat analogous to how binary numbers record word processing on computers or how symbols are used by court recorders.

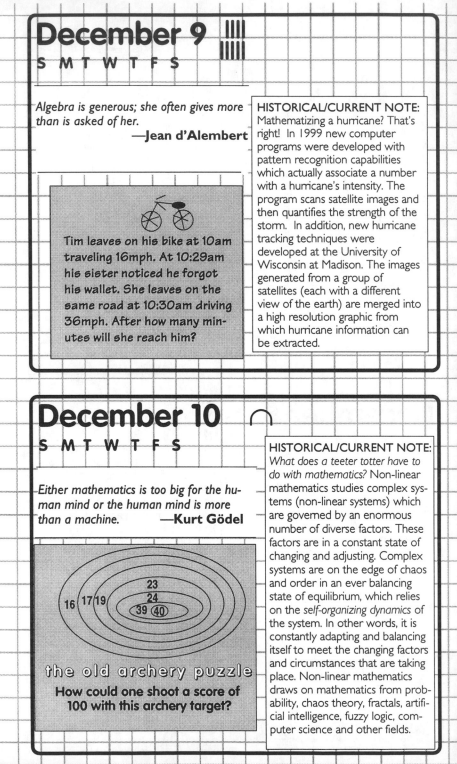

December 9

S M T W T F S

Algebra is generous; she often gives more than is asked of her.

—Jean d'Alembert

Tim leaves on his bike at 10am traveling 16mph. At 10:29am his sister noticed he forgot his wallet. She leaves on the same road at 10:30am driving 36mph. After how many minutes will she reach him?

HISTORICAL/CURRENT NOTE:
Mathematizing a hurricane? That's right! In 1999 new computer programs were developed with pattern recognition capabilities which actually associate a number with a hurricane's intensity. The program scans satellite images and then quantifies the strength of the storm. In addition, new hurricane tracking techniques were developed at the University of Wisconsin at Madison. The images generated from a group of satellites (each with a different view of the earth) are merged into a high resolution graphic from which hurricane information can be extracted.

December 10

S M T W T F S

Either mathematics is too big for the human mind or the human mind is more than a machine. **—Kurt Gödel**

16 17 19 23 24 39 40

the old archery puzzle

How could one shoot a score of 100 with this archery target?

HISTORICAL/CURRENT NOTE:
What does a teeter totter have to do with mathematics? Non-linear mathematics studies complex systems (non-linear systems) which are governed by an enormous number of diverse factors. These factors are in a constant state of changing and adjusting. Complex systems are on the edge of chaos and order in an ever balancing state of equilibrium, which relies on the *self-organizing dynamics* of the system. In other words, it is constantly adapting and balancing itself to meet the changing factors and circumstances that are taking place. Non-linear mathematics draws on mathematics from probability, chaos theory, fractals, artificial intelligence, fuzzy logic, computer science and other fields.

December 11 ∩I

S M T W T F S

Natural philosophy, mathematics and astronomy, carry the mind from the country to the creation, and give it fitness suited to the extent. —**Thomas Paine**

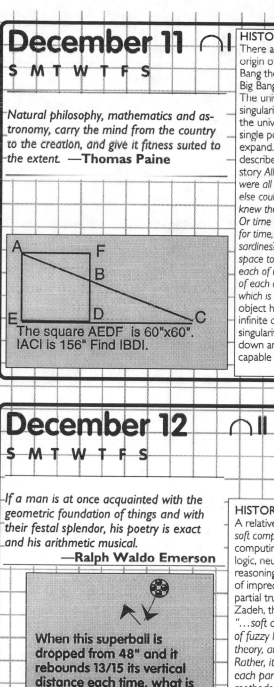

The square AEDF is 60"x60".
IACI is 156" Find IBDI.

HISTORICAL/CURRENT NOTE:
There are various theories about the origin of the universe. With the Big Bang theory, it is believed that the Big Bang created matter and space. The universe began at a point of singularity— when all the matter of the universe was concentrated in a single point, before it began to expand. Italian author Italo Calvino describes this point as follows in his story *All In One Point:* "*Naturally, we were all here—old Qfwfq said.—where else could we have been? Nobody knew then that there could be space. Or time either: what use did we have for time, packed in there like sardines?...in reality there wasn't even space to pack us into. Every point of each of us coincided with every point of each of the others in a single point, which is where we all were.*" Here an object has 0-dimension and mass of infinite density. At points of singularity, the laws of physics break down and there is no mathematics capable of describing them.

December 12 ∩II

S M T W T F S

If a man is at once acquainted with the geometric foundation of things and with their festal splendor, his poetry is exact and his arithmetic musical.
—**Ralph Waldo Emerson**

When this superball is dropped from 48" and it rebounds 13/15 its vertical distance each time, what is the distance in feet traveled in rebounds?

HISTORICAL/CURRENT NOTE:
A relatively new area of computing is *soft computing.* What is it? It's computing which relies on fuzzy logic, neural networks, probabilistic reasoning which deals with elements of imprecision, uncertainty and partial truths. As Professor Lotfi Zadeh, the creator of fuzzy sets, says "*...soft computing is not a melange of fuzzy logic, neural networks theory, and probabilistic reasoning. Rather, it is a partnership in which each partner contributes a distinct methodology for addressing problems ...*"

December 13

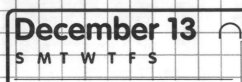

S M T W T F S

...the ancient geometers made use of a kind of analysis, which they employed in the solution of problems, although they begrudged to prosperity the knowledge of it.

—René Descartes

$$5 - 8 \times 3 \div \left(8 - 2\left(6 - 3\left(5 - 4\right) + 3\right)\right)$$

HISTORICAL/CURRENT NOTE:
Mathematics contests and prizes have prompted many discoveries and many non-valid proofs. During the 1500s there were many challenges and contests between mathematicians. Some were simply tests of abilities to solve a set problems in a public arena. These contests provided a forum for mathematicians to show off discoveries they had made in solving such problems as algebraic equations. Fermat's Last theorem prompted the French Academy of Sciences to offer a gold medal and 300 francs for its proof in 1815 and 1860. In 1909 Paul Wolfskehl bequeathed 100,000 marks for a published proof judged correct by the German Academy of Sciences. Within the first three years, over a thousand proofs were reported to have been submitted.

December 14

S M T W T F S

Mathematics is the queen of sciences and arithmetic is the queen of mathematics. She often condescends to render service to astronomy and other natural sciences, but under all circumstances the first place is her due.

—Karl Gauss

ABCD
- EFG
HI

E must be ___.

HISTORICAL/CURRENT NOTE:
For you mathematics aficionados who want to extend your repertoire of interesting mathematical constants beyond π, **e**, ϕ (the golden mean) check-out the *Euler-Mascheroni constant*—$\gamma \approx 0.577215664...$ which appears in number theory, *Catalan's constants*, which are associated with the numbers from the Pascal triangle, and the *Apéry's number*—
$$\zeta(3) = 1 + (1/2^3) + (1/3^3) + (1/4^3) + ...$$
$$\approx 1.20205...$$
These are just a few of the many constants that have popped up in such fields of mathematics as number theory, geometry, complex analysis. Once a significant constant has been found, then more math fun begins by trying to figure out whether it is algebraic, rational, irrational, and/or transcendental.

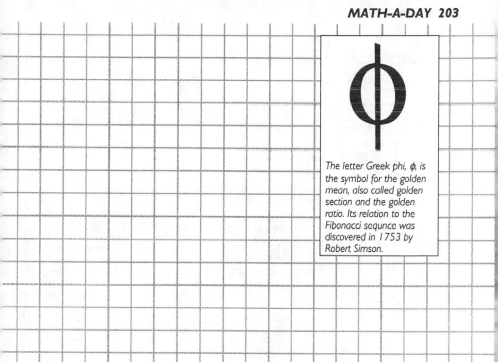

The letter Greek phi, φ, is the symbol for the golden mean, also called golden section and the golden ratio. Its relation to the Fibonacci sequnce was discovered in 1753 by Robert Simson.

December 15

S M T W T F S

Common sense is, as a matter of fact, nothing more than layers of preconceived notions stored in our memories and emotions for the most part before age eighteen. —**Albert Einstein**

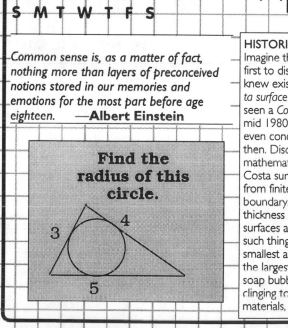

Find the radius of this circle.

3

4

5

HISTORICAL/CURRENT NOTE: Imagine the excitment of being the first to discover something no one knew existed. That's what the *Costa surface* was. No one had ever seen a *Costa surface* before the mid 1980s because no one had even conceived of one before then. Discovered by Brazilian mathematician Celso Costa, the Costa surface is a minimal surface from finite topology. It has no boundary, meaning it has no thickness and is infinite. Minimal surfaces are surfaces formed by such things as soap film where the smallest amount of film encases the largest amount of air, as in a soap bubble or as soap film clinging to a frame of various materials, such as wire.

December 16

S M T W T F S

What is it indeed that gives us the feeling of elegance in a solution…in a word it is all that introduces order; all that gives unity, that permits us to see clearly and to comprehend at once both the ensemble and the details. **—Henri Poincaré**

- Pick any three digits from the digits 0 through 9.
- Add 4 to the first choice, and multiply it by 10.
- Now add the second choice to this product, and multiply this by 10.
- To this add the third digit picked.
- Lastly, subtract 400.

How is it that the final number has all three digits you chose, and in the order you chose them?

HISTORICAL/CURRENT NOTE:
Mathematical moonshine? Mathematics has many examples where strange and crazy sounding math ideas have started out being considered foolish. The *Moonshine conjecture* seems to fit that category. In the 1970s British mathematicians John Conway and Simon Norton formulated the *Moonshine conjecture*, which relates two seemingly unrelated mathematical ideas— *elliptic function* and the *monster group*. Elliptic functions are used in many areas of mathematics and science, such as molecular structures of chemicals. *Monster group* is an enormous mathematical group with over 196,883 dimensions, countless symmetries, and more elements than there are particles in the universe. In August of 1998, mathematician Richard Borcherds received the Fields Medal for proving the *Moonshine conjecture* in 1989. In his proof he uses the ideas of string theory. Time will tell what new ground this mathematical moonshine will break.

December 17

S M T W T F S

The mathematician's best work is art, a high perfect art, as daring as the most secret dreams of imagination, clear and limpid. **—Gösta Mittag-Leffler**

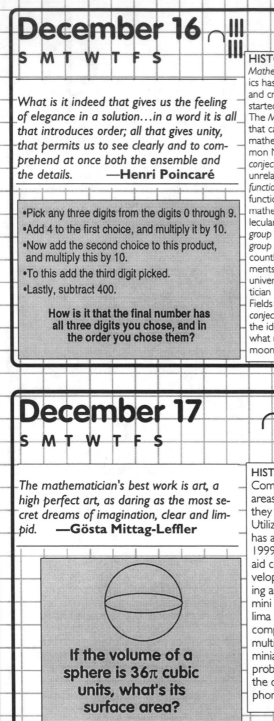

If the volume of a sphere is 36π cubic units, what's its surface area?

HISTORICAL/CURRENT NOTE:
Computers are invading so many areas, it's not surprising to learn they are now entering our ears. Utilizing the advances that science has achieved in miniaturization, in 1999 a number of Danish hearing aid companies announced the development of computerized hearing aids that fit in one's ears— mini twin computers the size of lima beams, each with its own computer operating system and multiple sound tracks. These new miniature audio systems will probably be used to also improve the quality of CD stereo, cellular phones, and Web audio links.

December 18
S M T W T F S

Arithmetic had entered the picture, with its many legs, its many spines and heads, its pitiless eyes made of zeros. Two and two made four, was its message. But what if you didn't have two and two? Then things wouldn't add up.
—Margaret Atwood,
The Blind Assassin

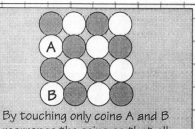

By touching only coins A and B rearrange the coins so that all columns feature the same shade.

HISTORICAL/CURRENT NOTE:
In the 1700s navigating the seas was the only practical means of carrying on international trade. To this end a number of countries posted prizes to entice people to come up with a reliable way of determining longitude. On July 8, 1714, England enacted the Longitude Act which established a Board of Longitude and a £20,000 first prize (≈$1,000,000 today) to the first person to devise a means for determining longitude to within an accuracy of 1/2 degree. Who would solve the longitude problem and claim the prize? Unfortunately, politics, intrigue, scandals and egos entered the picture. Many scientists, among them Isaac Newton, John Flamsteed, and Nevil Maskelyne, felt a mechanical device, such as a clock, would not be a reliable solution, and the only possible method lay with celestial calculations and mappings. John Flamsteed, the first astronomer royal at Greenwich, spent 40 years charting the night skies and compiling almanacs of tables and charts.

December 19
S M T W T F S

A philosopher once said "It is necessary for the very existence of science that the same conditions always produce the same results". Well, they do not.
—Richard Feynman

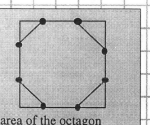

Find the area of the octagon formed from a 3x3 square. The vertices of the octagon are on points dividing the square's sides in thirds.

HISTORICAL/CURRENT NOTE:
In 1598, King Phillip III of Spain offered a lifetime pension as the prize for the solution to finding a reliable and feasible way of determining longitude. Galileo worked on this problem using the moons of Jupiter. Galileo's method required the sailor to observe Jupiter's moons and locate them on tables Galileo made. The difficulty of viewing these moons, especially aboard a ship, did not make this astronomical solution feasible

December 20

S M T W T F S

I am ill at these numbers.
—William Shakespeare
Hamlet

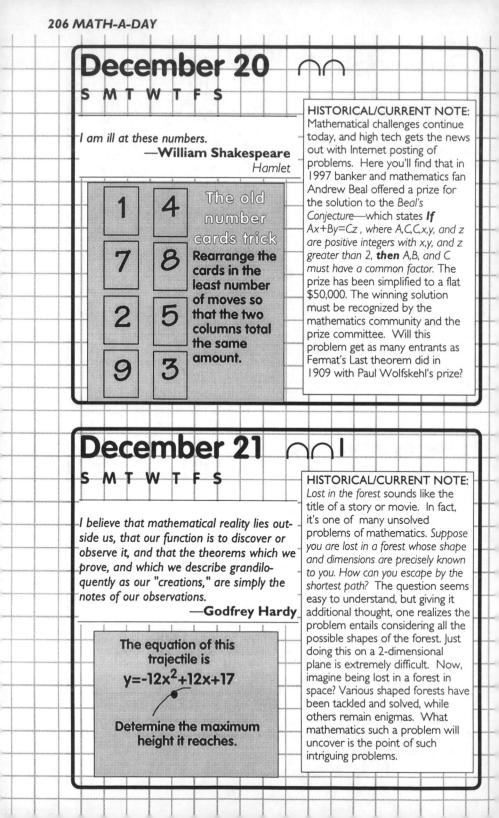

The old number cards trick

1 4
7 8
2 5
9 3

Rearrange the cards in the least number of moves so that the two columns total the same amount.

HISTORICAL/CURRENT NOTE:
Mathematical challenges continue today, and high tech gets the news out with Internet posting of problems. Here you'll find that in 1997 banker and mathematics fan Andrew Beal offered a prize for the solution to the *Beal's Conjecture*—which states **If** $Ax+By=Cz$, *where A,C,C,x,y, and z are positive integers with x,y, and z greater than 2,* **then** *A,B, and C must have a common factor.* The prize has been simplified to a flat $50,000. The winning solution must be recognized by the mathematics community and the prize committee. Will this problem get as many entrants as Fermat's Last theorem did in 1909 with Paul Wolfskehl's prize?

December 21

S M T W T F S

I believe that mathematical reality lies out-side us, that our function is to discover or observe it, and that the theorems which we prove, and which we describe grandilo-quently as our "creations," are simply the notes of our observations.
—Godfrey Hardy

The equation of this trajectile is
$$y=-12x^2+12x+17$$

Determine the maximum height it reaches.

HISTORICAL/CURRENT NOTE:
Lost in the forest sounds like the title of a story or movie. In fact, it's one of many unsolved problems of mathematics. *Suppose you are lost in a forest whose shape and dimensions are precisely known to you. How can you escape by the shortest path?* The question seems easy to understand, but giving it additional thought, one realizes the problem entails considering all the possible shapes of the forest. Just doing this on a 2-dimensional plane is extremely difficult. Now, imagine being lost in a forest in space? Various shaped forests have been tackled and solved, while others remain enigmas. What mathematics such a problem will uncover is the point of such intriguing problems.

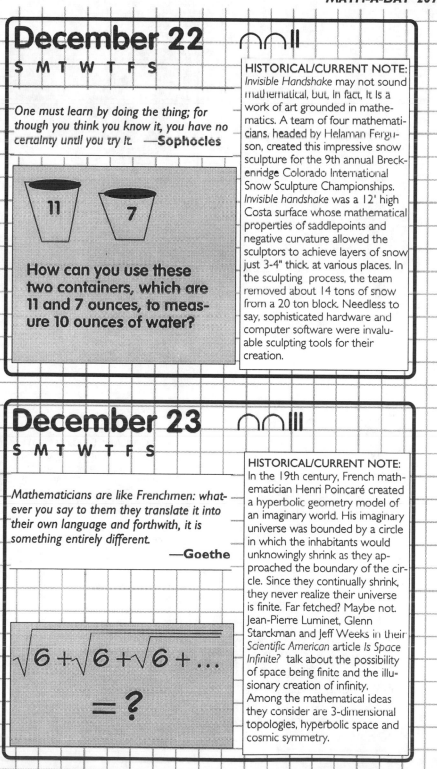

December 22

S M T W T F S

One must learn by doing the thing; for though you think you know it, you have no certainty until you try it. —**Sophocles**

11   7

How can you use these two containers, which are 11 and 7 ounces, to measure 10 ounces of water?

HISTORICAL/CURRENT NOTE:
Invisible Handshake may not sound mathematical, but, in fact, it is a work of art grounded in mathematics. A team of four mathematicians, headed by Helaman Ferguson, created this impressive snow sculpture for the 9th annual Breckenridge Colorado International Snow Sculpture Championships. *Invisible handshake* was a 12' high Costa surface whose mathematical properties of saddlepoints and negative curvature allowed the sculptors to achieve layers of snow just 3-4" thick. at various places. In the sculpting process, the team removed about 14 tons of snow from a 20 ton block. Needless to say, sophisticated hardware and computer software were invaluable sculpting tools for their creation.

December 23

S M T W T F S

Mathematicians are like Frenchmen: whatever you say to them they translate it into their own language and forthwith, it is something entirely different.
—**Goethe**

$$\sqrt{6+\sqrt{6+\sqrt{6+\ldots}}} = ?$$

HISTORICAL/CURRENT NOTE:
In the 19th century, French mathematician Henri Poincaré created a hyperbolic geometry model of an imaginary world. His imaginary universe was bounded by a circle in which the inhabitants would unknowingly shrink as they approached the boundary of the circle. Since they continually shrink, they never realize their universe is finite. Far fetched? Maybe not. Jean-Pierre Luminet, Glenn Starckman and Jeff Weeks in their *Scientific American* article *Is Space Infinite?* talk about the possibility of space being finite and the illusionary creation of infinity. Among the mathematical ideas they consider are 3-dimensional topologies, hyperbolic space and cosmic symmetry.

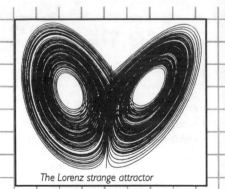

The Lorenz strange attractor

December 24
S M T W T F S

We have adroitly defined the infinite in arithmetic by a loveknot, in this manner ∞; but we possess not therefore the clearer notion of it. —**Voltaire**

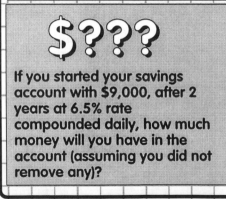

If you started your savings account with $9,000, after 2 years at 6.5% rate compounded daily, how much money will you have in the account (assuming you did not remove any)?

HISTORICAL/CURRENT NOTE:
What's a *strange attractor?* Mathematicians study patterns, and hidden in complex systems are patterns which illustrate the order within a chaotic system. Such a pattern is a *strange attractor.* The first *strange attractor* was discovered by meteorologist Edward Lorenz at MIT in 1962, and discussed in his ground breaking paper *Can the flap of a butterfly's wing stir a tornado in Texas?* This led to the popular phrase *butterfly effect*—a phenomenon in which minute changes in the initial conditions of a system can lead to enormous consequences. What systems display such patterns ? Everything from weather to economics. In fact, Adam Smith's invisible hand that steered the fluctuations of prices, goods, and services can be considered in this framework. Look at how prices of stocks, bonds, options, etc. in the financial market sometimes have enormous "adjustments" with no warning signs because of subtle changes taking place in the market.

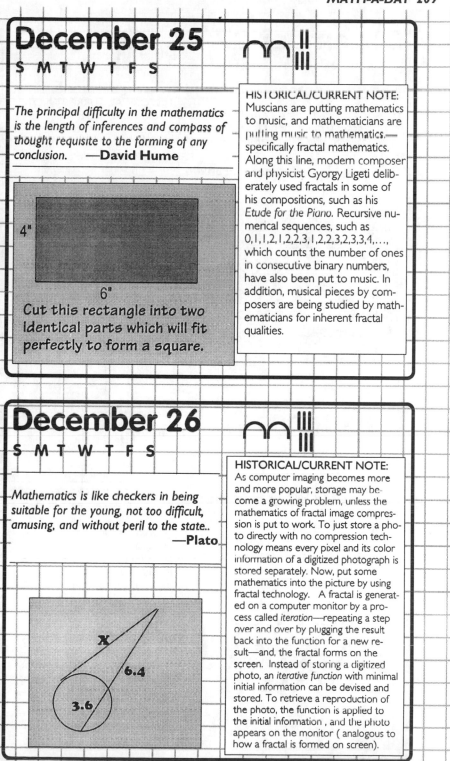

December 25
S M T W T F S

The principal difficulty in the mathematics is the length of inferences and compass of thought requisite to the forming of any conclusion. **—David Hume**

4"

6"

Cut this rectangle into two identical parts which will fit perfectly to form a square.

HISTORICAL/CURRENT NOTE: Muscians are putting mathematics to music, and mathematicians are putting music to mathematics.— specifically fractal mathematics. Along this line, modern composer and physicist Gyorgy Ligeti deliberately used fractals in some of his compositions, such as his *Etude for the Piano*. Recursive numerical sequences, such as 0,1,1,2,1,2,2,3,1,2,2,3,2,3,3,4,…, which counts the number of ones in consecutive binary numbers, have also been put to music. In addition, musical pieces by composers are being studied by mathematicians for inherent fractal qualities.

December 26
S M T W T F S

Mathematics is like checkers in being suitable for the young, not too difficult, amusing, and without peril to the state.. **—Plato**

X

6.4

3.6

HISTORICAL/CURRENT NOTE: As computer imaging becomes more and more popular, storage may become a growing problem, unless the mathematics of fractal image compression is put to work. To just store a photo directly with no compression technology means every pixel and its color information of a digitized photograph is stored separately. Now, put some mathematics into the picture by using fractal technology. A fractal is generated on a computer monitor by a process called *iteration*—repeating a step over and over by plugging the result back into the function for a new result—and, the fractal forms on the screen. Instead of storing a digitized photo, an *iterative function* with minimal initial information can be devised and stored. To retrieve a reproduction of the photo, the function is applied to the initial information , and the photo appears on the monitor (analogous to how a fractal is formed on screen).

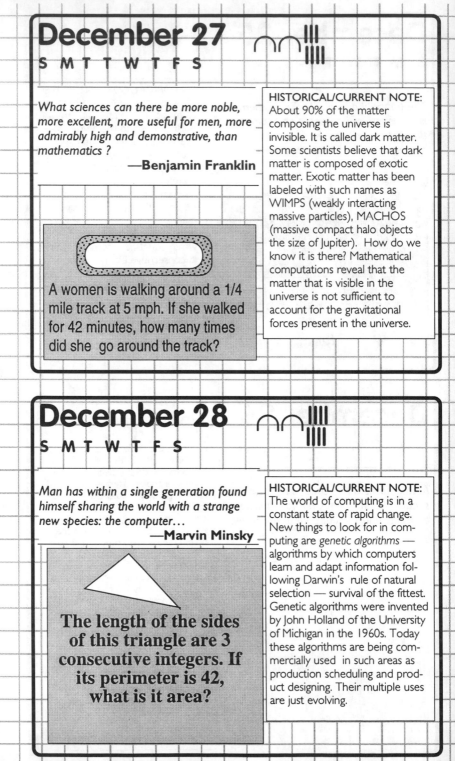

December 27
S M T T W T F S

What sciences can there be more noble, more excellent, more useful for men, more admirably high and demonstrative, than mathematics ?

—Benjamin Franklin

A women is walking around a 1/4 mile track at 5 mph. If she walked for 42 minutes, how many times did she go around the track?

HISTORICAL/CURRENT NOTE:
About 90% of the matter composing the universe is invisible. It is called dark matter. Some scientists believe that dark matter is composed of exotic matter. Exotic matter has been labeled with such names as WIMPS (weakly interacting massive particles), MACHOS (massive compact halo objects the size of Jupiter). How do we know it is there? Mathematical computations reveal that the matter that is visible in the universe is not sufficient to account for the gravitational forces present in the universe.

December 28
S M T T W T F S

Man has within a single generation found himself sharing the world with a strange new species: the computer...

—Marvin Minsky

The length of the sides of this triangle are 3 consecutive integers. If its perimeter is 42, what is it area?

HISTORICAL/CURRENT NOTE:
The world of computing is in a constant state of rapid change. New things to look for in computing are *genetic algorithms* — algorithms by which computers learn and adapt information following Darwin's rule of natural selection — survival of the fittest. Genetic algorithms were invented by John Holland of the University of Michigan in the 1960s. Today these algorithms are being commercially used in such areas as production scheduling and product designing. Their multiple uses are just evolving.

December 29

S M T W T F S

The most distinct and beautiful statement of any truth must take at last the mathematical form.

—**Henry Thoreau**

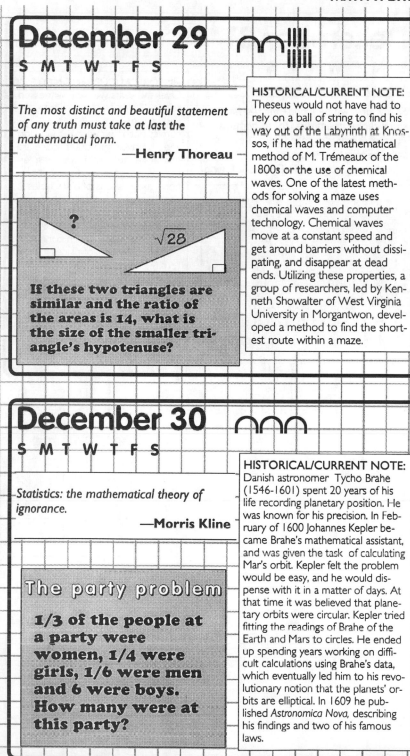

If these two triangles are similar and the ratio of the areas is 14, what is the size of the smaller triangle's hypotenuse?

HISTORICAL/CURRENT NOTE: Theseus would not have had to rely on a ball of string to find his way out of the Labyrinth at Knossos, if he had the mathematical method of M. Trémeaux of the 1800s or the use of chemical waves. One of the latest methods for solving a maze uses chemical waves and computer technology. Chemical waves move at a constant speed and get around barriers without dissipating, and disappear at dead ends. Utilizing these properties, a group of researchers, led by Kenneth Showalter of West Virginia University in Morgantwon, developed a method to find the shortest route within a maze.

December 30

S M T W T F S

Statistics: the mathematical theory of ignorance.

—**Morris Kline**

The party problem

1/3 of the people at a party were women, 1/4 were girls, 1/6 were men and 6 were boys. How many were at this party?

HISTORICAL/CURRENT NOTE: Danish astronomer Tycho Brahe (1546-1601) spent 20 years of his life recording planetary position. He was known for his precision. In February of 1600 Johannes Kepler became Brahe's mathematical assistant, and was given the task of calculating Mar's orbit. Kepler felt the problem would be easy, and he would dispense with it in a matter of days. At that time it was believed that planetary orbits were circular. Kepler tried fitting the readings of Brahe of the Earth and Mars to circles. He ended up spending years working on difficult calculations using Brahe's data, which eventually led him to his revolutionary notion that the planets' orbits are elliptical. In 1609 he published *Astronomica Nova*, describing his findings and two of his famous laws.

December 31 ∩∩∩∣

S M T W T F S

So if man's wit be wandering, let him study the mathematics: for in demonstrations, if his wit be called away never so little, he must begin again.
—**Sir Francis Bacon**

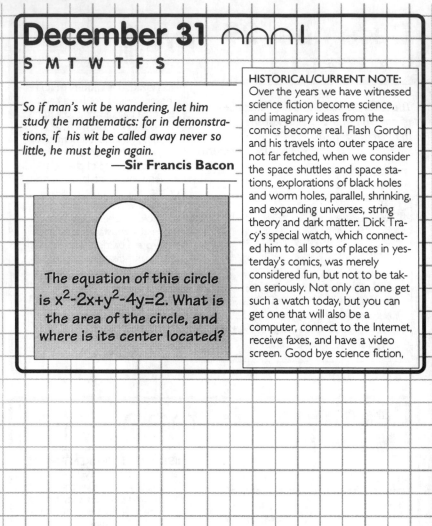

The equation of this circle is $x^2-2x+y^2-4y=2$. What is the area of the circle, and where is its center located?

HISTORICAL/CURRENT NOTE:
Over the years we have witnessed science fiction become science, and imaginary ideas from the comics become real. Flash Gordon and his travels into outer space are not far fetched, when we consider the space shuttles and space stations, explorations of black holes and worm holes, parallel, shrinking, and expanding universes, string theory and dark matter. Dick Tracy's special watch, which connected him to all sorts of places in yesterday's comics, was merely considered fun, but not to be taken seriously. Not only can one get such a watch today, but you can get one that will also be a computer, connect to the Internet, receive faxes, and have a video screen. Good bye science fiction,

°The subtraction symbol may have orignated in Medieval times. The traders of this period used "–" to indicate the differences in the weights of items.

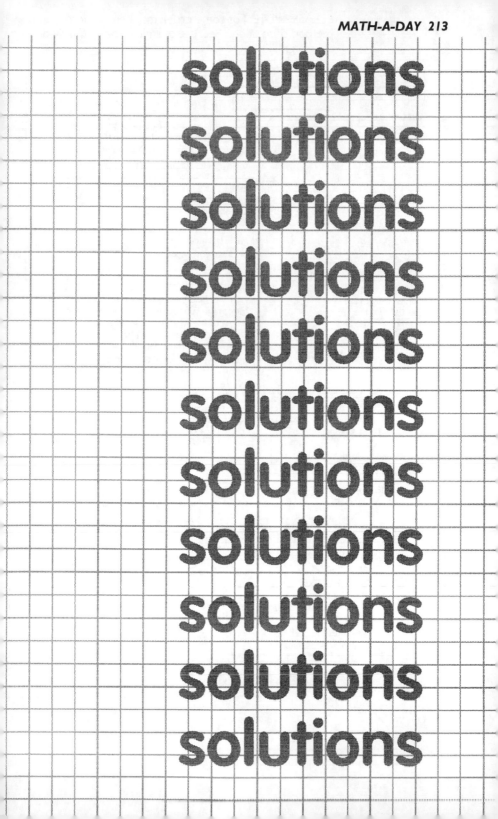

PLEASE NOTE: For some problems, there is more than one method of solution. Because of limited space, usually only one solution is illustrated for each problem.

JAN 1 The can makes a 12x4 retangle when unfolded.

4 [12   ]

circumference is 12=diameter•π

—>d=12/π—> radius=6/π.—>

$\text{Volume}_{cyl.} = \left[\left(\frac{6}{\pi}\right)^2 \pi\right] \cdot 4 = \frac{144}{\pi}$

JAN 2
Fill up 3 qt. jar, and pour it into the 5 qt. jar. Refill 3 qt. jar, and fill up the 5 qt. with it. This leaves 1 qt. in the 3 qt. jar. Pour out the 5 qt. water, and pour into it the 1 qt. from the 3 qt. jar. Refill the 3 qt. jar. Pour it into the 5 qt. jar. This makes 4 qt.s in the 5 qt. jar.

JAN 3

$(1+2)^2 + (y-5)^2 + (-3+4)^2 = 14$

$y^2 - 10y + 21 = 0$

$(y-7)(y-3)$

$y = 7$

JAN 4 Filling in the angles measurement, the large triangle comes out to be an isosceles right triangle. Call each of its legs x, then

$x^2 + x^2 = (\sqrt{128})^2$ by the Pythagorean theorem

$2x^2 = 128 \longrightarrow x^2 = 64$

$x = 8$, taking only the positive root

Using a the 30°-60°-90° right triangle theorem, |PQ|=4.

JAN 5 The small and large triangles are similar because they share a common angle and both have an angle of a°. Therefore their corresponding sides are proportional.

$\frac{5}{8} = \frac{6\frac{7}{8}}{x}$

$5x = 55$

$x = 11$

JAN 6 Find the area of the 4 semicircles and from this subtract the difference between the area of the white circle and the square. Each semicircle's area is (1/2)(4π). All together, 4•[(1/2)(4π)]=8π. The diameter of the white circle is the diagonal of the square, 4√2, making its radius 2√2, so its area is =8π. The square's area is 16.
—> the shaded area is 8π −(8π −16)=16.

JAN 7

$\sum_{n=1}^{21} 1 = 1+1+1+\ldots+1 = 21$

21 ones added together.

JAN 8
Changing 32 to base 8 means the 3 is worth 3 eights and 2 is 2 ones which equals 26 in base ten.

JAN 9

$\sqrt[18]{27^6} = 27^{\frac{6}{18}} = 27^{\frac{1}{3}} = 3$

JAN 10 A circle is 2π radians which is roughly 2(3.14...)≈6

JAN 11 Any number raised to the 0 power is 1, except 0, which is undefined. So this expression comes out ot be 1.

JAN 12

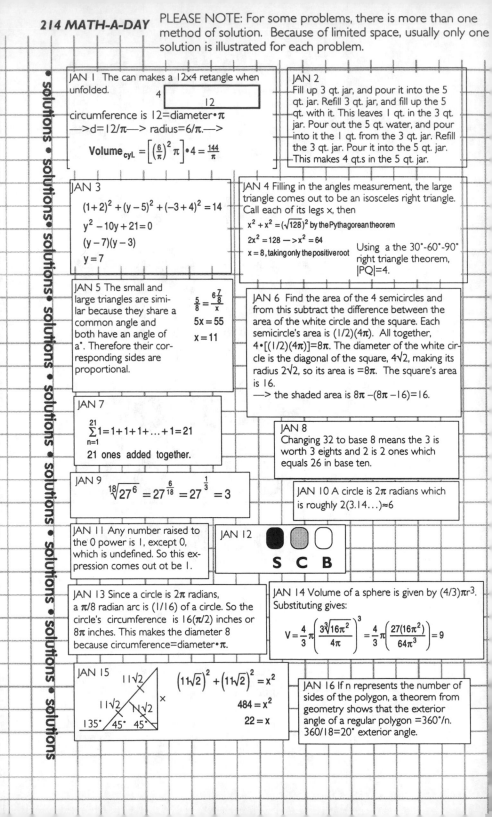

S   C   B

JAN 13 Since a circle is 2π radians, a π/8 radian arc is (1/16) of a circle. So the circle's circumference is 16(π/2) inches or 8π inches. This makes the diameter 8 because circumference=diameter•π.

JAN 14 Volume of a sphere is given by (4/3)πr³. Substituting gives:

$V = \frac{4}{3}\pi\left(\frac{3\sqrt[3]{16\pi^2}}{4\pi}\right)^3 = \frac{4}{3}\pi\left(\frac{27(16\pi^2)}{64\pi^3}\right) = 9$

JAN 15

11√2, 11√2, N√2, x, 135°, 45°, 45°

$\left(11\sqrt{2}\right)^2 + \left(11\sqrt{2}\right)^2 = x^2$

$484 = x^2$

$22 = x$

JAN 16 If n represents the number of sides of the polygon, a theorem from geometry shows that the exterior angle of a regular polygon =360°/n. 360/18=20° exterior angle.

JAN 17 Assuming 365 days per year. A represents accummulated interest, P represents initial principal, r= interest rate, n=number of times compounded yearly, t=the number of years. The extra day from a leap year would not significantly change the answer.

$$A = P\left(1 + \tfrac{r}{n}\right)^{nt} = \$20\left(1 + \frac{.05}{365}\right)^{365 \cdot 6} \approx \$27$$

JAN 18

$$-\tfrac{1}{2}x + 0.6(x + 5) = 4.4$$
$$-0.5x + 0.6x + 3 = 4.4$$
$$0.1x = 1.4$$
$$x = 14$$

JAN 19 We need to find how far a car going 10mph will have gone in 5 minutes. 5 minutes which is 5/60 of an hour.
(5/60)hr.•10mph
= (5/6) miles
=(5/6) 5280'≈4400'.

(Note:
5280'=1 mile)

JAN 20 Ask, "Point to the path that leads to the city you come from." Regardless whether he is from the city of Truths or Lies, he will point to the path to the city of Truths.

JAN 21 The two terms differ by 1200.

$$a_1 = 6 \cdot 1 - 1 = 5$$
$$a_{201} = 6(201) - 1 = 1205$$

JAN 22

$$y = 2x^3 - 7x + 3$$
$$y' = 6x^2 - 7$$
$$\text{at } x = 2, \ y' = 6 \cdot 2^2 - 7 = 17$$

JAN 23

$$\sqrt{6+x} \geq 4 \ \cap \ 7x - 10 \leq 6x$$
$$6 + x \geq 16 \ \cap \ 1x \leq 10$$
$$x \geq 10 \ \cap \ x \leq 10 \rightarrow x = 10$$

JAN 24 $x + 2y = 19$ $\boxed{x=29; \ y=-5; \ z=17}$
$x = y + 2z$ these give → $y + 2z + 2y = 19 \rightarrow 3y + 2z = 19$ (a)
Using $3z = 2x - 7$ and $x = y + 2z$ gives $3z = 2(y + 2z) - 7 \rightarrow$
$-2y - z = -7$ (b)
Using (a) & (b)
$3y + 2z = 19 \rightarrow 3y + 2z = 19$
$-2y - z = -7 \rightarrow -4y - 2z = -14 \rightarrow -y = 5 \rightarrow y = -5$
$10 - z = -7 \rightarrow z = 17$
Using $x = y + 2z \ \rightarrow x = -5 + 2 \cdot 17 = 29$

JAN 25 Counting the 12 diagonals on all 6 faces plus the 4 interior diagonals, there are 16.

JAN 26 $(\overline{.142857}) = \tfrac{1}{7} \rightarrow \tfrac{1}{7} \cdot 14 = 2$

JAN 27 With his solution of the Könisgberg bridge problem, Euler showed that if a network had more than 2 odd vertices it was not traceable. A vertex is odd if it has an odd number of lines segments through it. This network is not traceable. It has 4 odd vertices, three with 3 segments and one with 5 segments.

JAN 28 A quadratic equation has only two solutions. If one of them is complex and the coefficients are rational numbers, the other solution must be the conjugate of the original solution. Multiplying out the equation: (x-(1-3i))(x-(1+3i))=0, gives the constant term. An easiesr way to answer this question is to remember the constant term comes from the product of the two roots, namely (1+3i)(1-3i)=10.

JAN 29 E=1. Even if A and C were the highest digits, namely 9 & 8, there sum would be 17 and you would not be carrying more than 1 from the sum of B & C, and that would make 18, meaning E must be 1.

JAN 31 The volume of a pyramid is given by taking 1/3 the area of its base times its altitude. Its base is an equilateral triangle with sides 8. A perpendicular bisector of an equilateral triangle bisects the opposite side and forms a 30°-60°-90° right triangle.

Its area is (1/2) 8(4√3)=16√3

—> volume of the pyramid
=(1/3)(16√3)√3
=16 cubic units

JAN 30

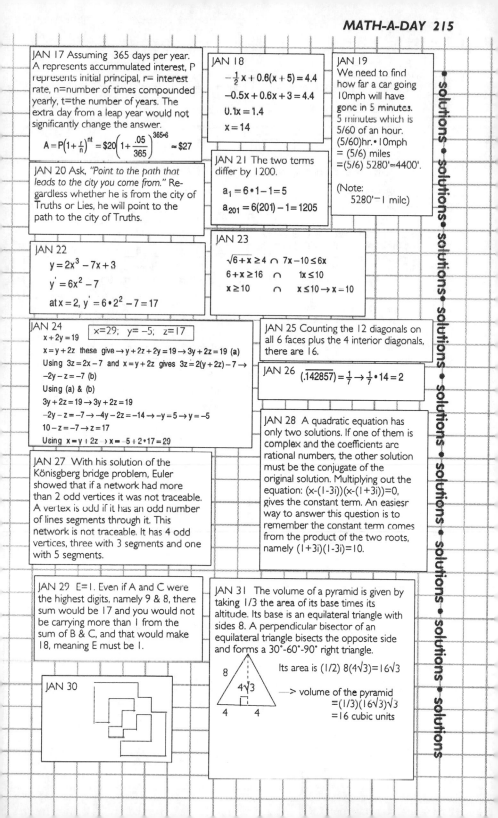

FEB 1 When midpoints of two sides of a triangle are joined, the resulting segment is parallel to the third side and have its length. Thus,
$$|AB| = \text{half of } |PQ| = 23.$$

FEB 3 In 1658, Christopher Wren discovered that the length of one cycle of a cycloid is 4 times the circle's diameter.
—>4(6.5)=26

FEB 4 An angle inscribed in a circle has measure 1/2 the arc it intercepts. The arc for angle x is 180°-124°=56° because of the semicircle.
So x°=(1/2)56°=28°.

FEB 6
$$f(12) = \sqrt{3 \cdot 12} = 6$$
$$g(6) = 6^2 - 5 \cdot 6 - 1 = 5$$

FEB 9   F = 71.6° and since
$$F = \tfrac{9}{5}C + 32 \rightarrow$$
$$71.6 = \tfrac{9}{5}C + 32$$
$$39.6° = \tfrac{9}{5}C \rightarrow 22° = C$$

FEB 10
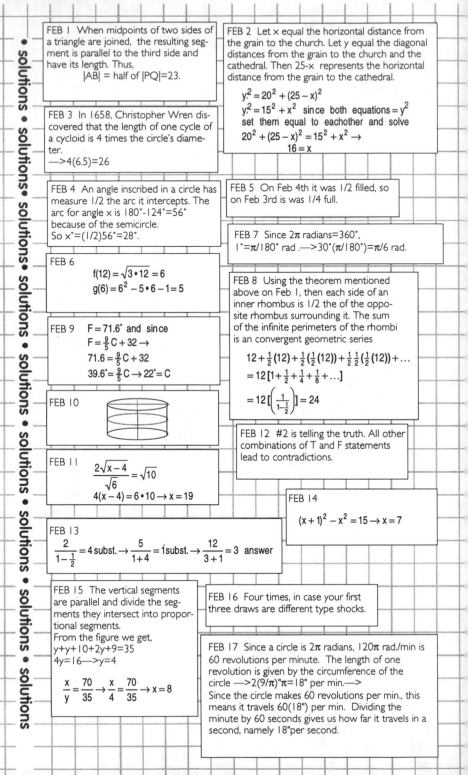

FEB 11
$$\frac{2\sqrt{x-4}}{\sqrt{6}} = \sqrt{10}$$
$$4(x-4) = 6 \cdot 10 \rightarrow x = 19$$

FEB 13
$$\frac{2}{1-\tfrac{1}{2}} = 4 \text{ subst.} \rightarrow \frac{5}{1+4} = 1 \text{ subst.} \rightarrow \frac{12}{3+1} = 3 \text{ answer}$$

FEB 15 The vertical segments are parallel and divide the segments they intersect into proportional segments.
From the figure we get,
y+y+10+2y+9=35
4y=16—>y=4
$$\frac{x}{y} = \frac{70}{35} \rightarrow \frac{x}{4} = \frac{70}{35} \rightarrow x = 8$$

FEB 2 Let x equal the horizontal distance from the grain to the church. Let y equal the diagonal distances from the grain to the church and the cathedral. Then 25-x represents the horizontal distance from the grain to the cathedral.
$$y^2 = 20^2 + (25-x)^2$$
$$y^2 = 15^2 + x^2 \quad \text{since both equations} = y^2$$
set them equal to eachother and solve
$$20^2 + (25-x)^2 = 15^2 + x^2 \rightarrow$$
$$16 = x$$

FEB 5 On Feb 4th it was 1/2 filled, so on Feb 3rd is was 1/4 full.

FEB 7 Since 2π radians=360°,
1°=π/180° rad.—>30°(π/180°)=π/6 rad.

FEB 8 Using the theorem mentioned above on Feb 1, then each side of an inner rhombus is 1/2 the of the opposite rhombus surrounding it. The sum of the infinite perimeters of the rhombi is an convergent geometric series
$$12 + \tfrac{1}{2}(12) + \tfrac{1}{2}(\tfrac{1}{2}(12)) + \tfrac{1}{2}\tfrac{1}{2}(\tfrac{1}{2}(12)) + \ldots$$
$$= 12[1 + \tfrac{1}{2} + \tfrac{1}{4} + \tfrac{1}{8} + \ldots]$$
$$= 12\left[\left(\frac{1}{1-\tfrac{1}{2}}\right)\right] = 24$$

FEB 12 #2 is telling the truth. All other combinations of T and F statements lead to contradictions.

FEB 14
$$(x+1)^2 - x^2 = 15 \rightarrow x = 7$$

FEB 16 Four times, in case your first three draws are different type shocks.

FEB 17 Since a circle is 2π radians, 120π rad./min is 60 revolutions per minute. The length of one revolution is given by the circumference of the circle —>2(9/π)"π=18" per min.—>
Since the circle makes 60 revolutions per min., this means it travels 60(18") per min. Dividing the minute by 60 seconds gives us how far it travels in a second, namely 18"per second.

FEB 18 The number of diagonals per convex n-gon can be derived, and is $(n^2-3n)/2$. So 65-gon would have $(65^2-3\cdot65)/2 \longrightarrow 2015$

FEB 19 \$7,300,000/1000 --> 7,300 days --> 20 years.

FEB 22 Let $x°$ represent angle A's measure, then

$$180 - x° = 3(90 - x°) \to x = 45°$$

FEB 20 $(1/5)\dot{y}=5\sin2\Theta \longrightarrow$
$y=25\sin2\Theta \longrightarrow$
amplitude is 25

FEB 21 Solving these two equations simultaneously, we get

$$\frac{54+x+y+43+21}{5} = 40.4 \quad \text{and} \quad \frac{x}{y}=6 \quad \to y=12 \ \& \ x=72$$

FEB 24 Tom and Jerry are baseball players on opposing teams.

FEB 23 By circumscribing the star by a pentagon, you can arrive at the measure of each point to be 36°.

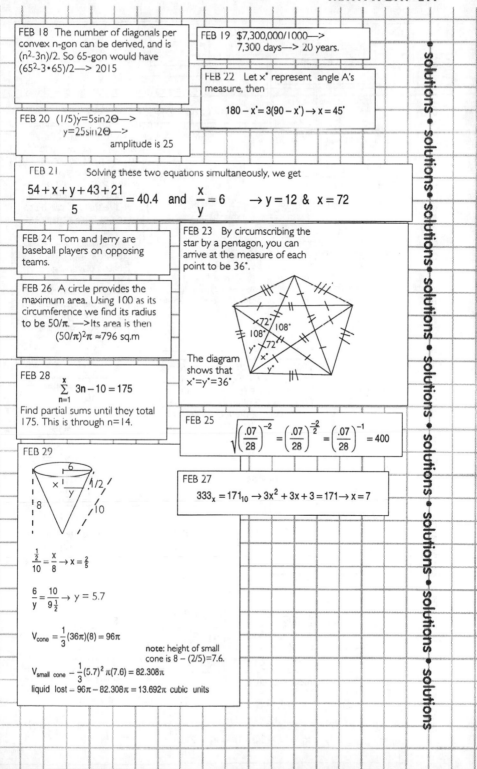

The diagram shows that $x°=y°=36°$

FEB 26 A circle provides the maximum area. Using 100 as its circumference we find its radius to be $50/\pi$. —>Its area is then $(50/\pi)^2\pi \approx 796$ sq.m

FEB 28 $\sum\limits_{n=1}^{x} 3n-10 = 175$

Find partial sums until they total 175. This is through n=14.

FEB 25 $$\sqrt{\left(\frac{.07}{28}\right)^{-2}} = \left(\frac{.07}{28}\right)^{\frac{-2}{2}} = \left(\frac{.07}{28}\right)^{-1} = 400$$

FEB 29

FEB 27
$$333_x = 171_{10} \to 3x^2+3x+3 = 171 \to x = 7$$

$$\frac{\frac{1}{2}}{10} = \frac{x}{8} \to x = \tfrac{2}{5}$$

$$\frac{6}{y} = \frac{10}{9\frac{1}{2}} \to y = 5.7$$

$$V_{cone} = \frac{1}{3}(36\pi)(8) = 96\pi$$

note: height of small cone is $8 - (2/5)=7.6$.

$$V_{small\ cone} = \frac{1}{3}(5.7)^2\,\pi(7.6) = 82.308\pi$$

liquid lost $= 96\pi-82.308\pi = 13.692\pi$ cubic units

MAR 1
$$-1-9^{\frac{1}{2}}+8^{\frac{1}{3}}+9^{\frac{1}{2}}$$
$$=-1-3+2+3=1$$

MAR 2

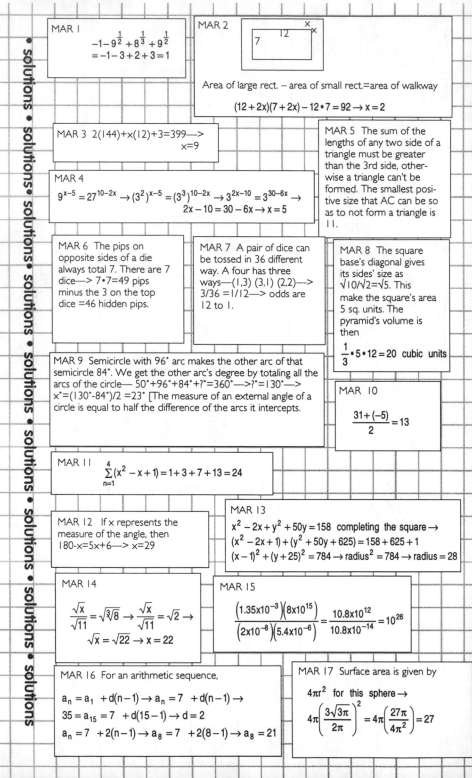

Area of large rect. − area of small rect.=area of walkway

$$(12+2x)(7+2x)-12\bullet 7=92 \rightarrow x=2$$

MAR 3 $2(144)+x(12)+3=399\longrightarrow$
$$x=9$$

MAR 5 The sum of the lengths of any two side of a triangle must be greater than the 3rd side, otherwise a triangle can't be formed. The smallest positive size that AC can be so as to not form a triangle is 11.

MAR 4
$$9^{x-5}=27^{10-2x} \rightarrow (3^2)^{x-5}=(3^3)^{10-2x} \rightarrow 3^{2x-10}=3^{30-6x} \rightarrow$$
$$2x-10=30-6x \rightarrow x=5$$

MAR 6 The pips on opposite sides of a die always total 7. There are 7 dice—> $7\bullet 7=49$ pips minus the 3 on the top dice $=46$ hidden pips.

MAR 7 A pair of dice can be tossed in 36 different way. A four has three ways—(1,3) (3,1) (2,2)—> $3/36 =1/12$—> odds are 12 to 1.

MAR 8 The square base's diagonal gives its sides' size as $\sqrt{10}/\sqrt{2}=\sqrt{5}$. This make the square's area 5 sq. units. The pyramid's volume is then
$$\frac{1}{3}\bullet 5\bullet 12=20 \text{ cubic units}$$

MAR 9 Semicircle with 96° arc makes the other arc of that semicircle 84°. We get the other arc's degree by totaling all the arcs of the circle— $50°+96°+84°+?°=360°$—>?°$=130°$—> $x°=(130°-84°)/2 =23°$ [The measure of an external angle of a circle is equal to half the difference of the arcs it intercepts.

MAR 10
$$\frac{31+(-5)}{2}=13$$

MAR 11
$$\sum_{n=1}^{4}(x^2 -x+1)=1+3+7+13=24$$

MAR 12 If x represents the measure of the angle, then $180-x=5x+6$—> $x=29$

MAR 13
$$x^2 -2x+y^2 +50y =158 \text{ completing the square} \rightarrow$$
$$(x^2 -2x+1)+(y^2 +50y +625)=158+625+1$$
$$(x-1)^2 +(y+25)^2 =784 \rightarrow \text{radius}^2 =784 \rightarrow \text{radius}=28$$

MAR 14
$$\frac{\sqrt{x}}{\sqrt{11}}=\sqrt[3]{\sqrt{8}} \rightarrow \frac{\sqrt{x}}{\sqrt{11}}=\sqrt{2} \rightarrow$$
$$\sqrt{x}=\sqrt{22} \rightarrow x=22$$

MAR 15
$$\frac{(1.35\times10^{-3})(8\times10^{15})}{(2\times10^{-8})(5.4\times10^{-6})}=\frac{10.8\times10^{12}}{10.8\times10^{-14}}=10^{26}$$

MAR 16 For an arithmetic sequence,
$$a_n = a_1 +d(n-1) \rightarrow a_n =7 +d(n-1) \rightarrow$$
$$35 = a_{15} =7 +d(15-1) \rightarrow d=2$$
$$a_n =7 +2(n-1) \rightarrow a_8 =7 +2(8-1) \rightarrow a_8 =21$$

MAR 17 Surface area is given by $4\pi r^2$ for this sphere $\rightarrow$
$$4\pi\left(\frac{3\sqrt{3\pi}}{2\pi}\right)^2 =4\pi\left(\frac{27\pi}{4\pi^2}\right)=27$$

MAR 18   Since $e^{\pi i} = -1$,

$$\frac{-e^{\pi i}}{\log_2(\log_8 64)} = \frac{-(-1)}{\log_2(2)} = \frac{1}{1} = 1$$

MAR 19 Empty the 2nd glass from the left into the the 2nd glass from the right, and then replace the glass to its original position.

MAR 20 The 3 triangles pictured are similar triangles since all have a right angle and share a common angle. Using proportional parts of similar triangles, we get

$$\frac{15}{x} = \frac{x+16}{15} \rightarrow$$

Note:
△ABD~△BCD~△ACB
Let |AD|=x &
x+16=|AC|.

$x = 9 \rightarrow |AC| = 9 + 16 = 25$

MAR 21

$f(x) = 3 + \sqrt{x-6}$
$f^{-1}: \ x = 3 + \sqrt{y-6}$
$f^{-1}(8) \ \ 8 = 3 + \sqrt{y-6}$
$\qquad 5 = \sqrt{y-6} \rightarrow y = 31$

MAR 22

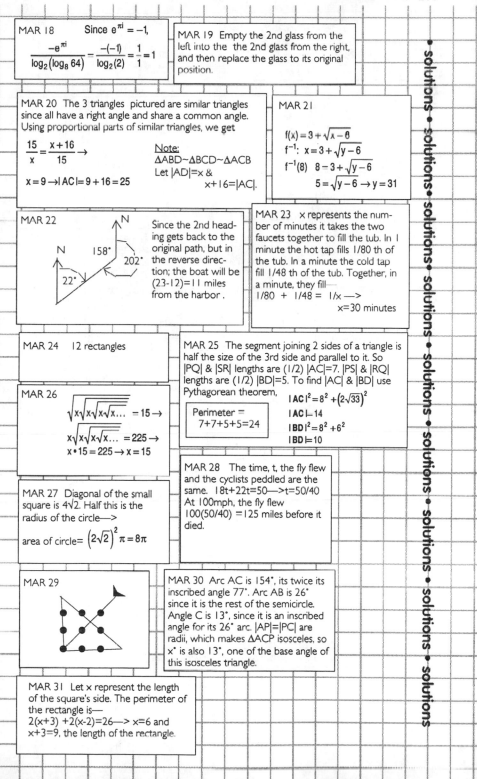

Since the 2nd heading gets back to the original path, but in the reverse direction; the boat will be (23-12)=11 miles from the harbor.

MAR 23 x represents the number of minutes it takes the two faucets together to fill the tub. In 1 minute the hot tap fills 1/80 th of the tub. In a minute the cold tap fill 1/48 th of the tub. Together, in a minute, they fill—
$1/80 + 1/48 = 1/x \rightarrow$
$\qquad\qquad x = 30$ minutes

MAR 24   12 rectangles

MAR 25 The segment joining 2 sides of a triangle is half the size of the 3rd side and parallel to it. So |PQ| & |SR| lengths are (1/2) |AC|=7. |PS| & |RQ| lengths are (1/2) |BD|=5. To find |AC| & |BD| use Pythagorean theorem,

Perimeter =
7+7+5+5=24

$|AC|^2 = 8^2 + (2\sqrt{33})^2$
$|AC| = 14$
$|BD|^2 = 8^2 + 6^2$
$|BD| = 10$

MAR 26

$$\sqrt{x\sqrt{x\sqrt{x\sqrt{x}\ldots}}} = 15 \rightarrow$$

$$x\sqrt{x\sqrt{x\sqrt{x}\ldots}} = 225 \rightarrow$$
$$x \cdot 15 = 225 \rightarrow x = 15$$

MAR 27 Diagonal of the small square is $4\sqrt{2}$. Half this is the radius of the circle—>

area of circle= $\left(2\sqrt{2}\right)^2 \pi = 8\pi$

MAR 28 The time, t, the fly flew and the cyclists peddled are the same. 18t+22t=50—>t=50/40 At 100mph, the fly flew 100(50/40) =125 miles before it died.

MAR 29

MAR 30 Arc AC is 154°, its twice its inscribed angle 77°. Arc AB is 26° since it is the rest of the semicircle. Angle C is 13°, since it is an inscribed angle for its 26° arc. |AP|=|PC| are radii, which makes △ACP isosceles, so x° is also 13°, one of the base angle of this isosceles triangle.

MAR 31 Let x represent the length of the square's side. The perimeter of the rectangle is—
2(x+3) +2(x-2)=26—> x=6 and
x+3=9, the length of the rectangle.

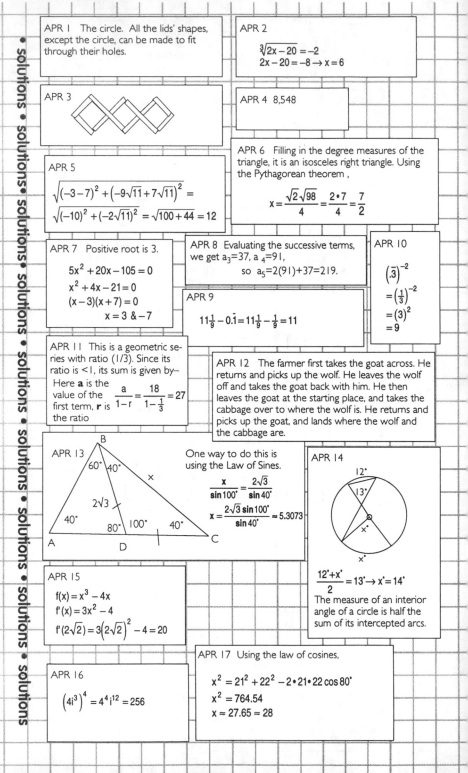

APR 1 The circle. All the lids' shapes, except the circle, can be made to fit through their holes.

APR 2
$$\sqrt[3]{2x-20}=-2$$
$$2x-20=-8 \rightarrow x=6$$

APR 3

APR 4 8,548

APR 5
$$\sqrt{(-3-7)^2+\left(-9\sqrt{11}+7\sqrt{11}\right)^2}=$$
$$\sqrt{(-10)^2+(-2\sqrt{11})^2}=\sqrt{100+44}=12$$

APR 6 Filling in the degree measures of the triangle, it is an isosceles right triangle. Using the Pythagorean theorem ,
$$x=\frac{\sqrt{2}\sqrt{98}}{4}=\frac{2\cdot7}{4}=\frac{7}{2}$$

APR 7 Positive root is 3.
$$5x^2+20x-105=0$$
$$x^2+4x-21=0$$
$$(x-3)(x+7)=0$$
$$x=3\ \&-7$$

APR 8 Evaluating the successive terms, we get $a_3=37$, $a_4=91$, so $a_5=2(91)+37=219$.

APR 9
$$11\tfrac{1}{9}-0.\bar{1}=11\tfrac{1}{9}-\tfrac{1}{9}=11$$

APR 10
$$(.\bar{3})^{-2}$$
$$=\left(\tfrac{1}{3}\right)^{-2}$$
$$=(3)^2$$
$$=9$$

APR 11 This is a geometric series with ratio (1/3). Since its ratio is <1, its sum is given by— Here **a** is the value of the first term, **r** is the ratio
$$\frac{a}{1-r}=\frac{18}{1-\frac{1}{3}}=27$$

APR 12 The farmer first takes the goat across. He returns and picks up the wolf. He leaves the wolf off and takes the goat back with him. He then leaves the goat at the starting place, and takes the cabbage over to where the wolf is. He returns and picks up the goat, and lands where the wolf and the cabbage are.

APR 13
60° 40° B, x, 2√3, 40° A, 80° 100° D, 40° C

One way to do this is using the Law of Sines.
$$\frac{x}{\sin100°}=\frac{2\sqrt{3}}{\sin40°}$$
$$x=\frac{2\sqrt{3}\sin100°}{\sin40°}\approx5.3073$$

APR 14
12°, 13°, x°, x°
$$\frac{12°+x°}{2}=13° \rightarrow x'=14°$$
The measure of an interior angle of a circle is half the sum of its intercepted arcs.

APR 15
$$f(x)=x^3-4x$$
$$f'(x)=3x^2-4$$
$$f'(2\sqrt{2})=3\left(2\sqrt{2}\right)^2-4=20$$

APR 16
$$\left(4i^3\right)^4=4^4i^{12}=256$$

APR 17 Using the law of cosines,
$$x^2=21^2+22^2-2\cdot21\cdot22\cos80°$$
$$x^2=764.54$$
$$x\approx27.65\approx28$$

APR 18

1	2	3	4	5	6	7
●	●	●		○	◐	○

3B moves to square 4.
5G to 3. 6G to 5. 4B to 6. 2B to 4.
1B to 2. 3G to 1. 5G to 3. 7G to 5.
6B to 7. 4B to 6. 2B to 4. 3G to 2.
5G to 3. 4B to 5.

APR 20 $\log_x 1,000,000 = 6$
$\log_x 10^6 = 6$
$x^6 = 10^6$
$x = 10$

APR 22 Babylonian base 60

$20(60) + 3 = 1203$

APR 23

$x^2 - 4x + 169y^2 = 165$ complete the square
$(x^2 - 4x + 4) + 169y^2 = 165 + 4$
$(x-2)^2 + 169y^2 = 169$
$\dfrac{(x-2)^2}{169} + \dfrac{y^2}{1} = 1 \rightarrow$ major axis $= 2\sqrt{169}$
$= 26$

APR 25

$\dfrac{(3.2 \times 10^5)(2 \times 10^8)}{20\left(4 \times 10^{-4}\right)(5)} = 1.6 \times 10^{15}$

APR 27 $y = 5\sin(2\theta + 90°)$
$y = 5\sin 2(\theta + 45°)$

APR 29

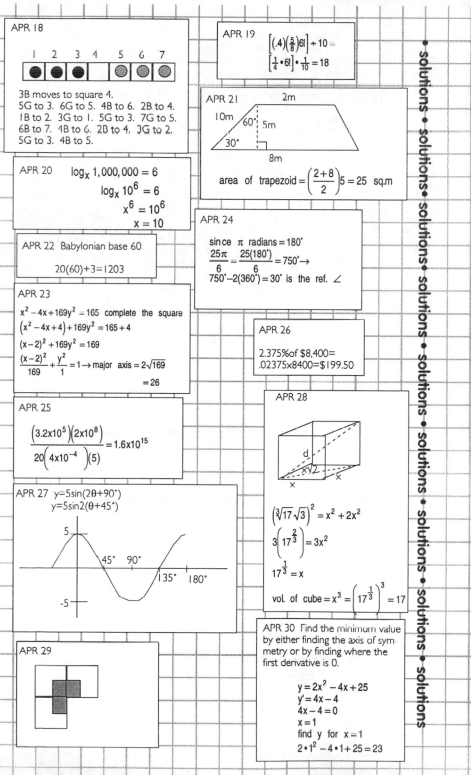

APR 19 $\left[(.4)\left(\frac{5}{8}\right)6!\right] + 10 =$
$\left[\frac{1}{4} \cdot 6!\right] \cdot \frac{1}{10} = 18$

APR 21

area of trapezoid $= \left(\dfrac{2+8}{2}\right)5 = 25$ sq.m

APR 24

since π radians $= 180°$
$\dfrac{25\pi}{6} = \dfrac{25(180°)}{6} = 750° \rightarrow$
$750° - 2(360°) = 30°$ is the ref. $\angle$

APR 26

2.375% of $8,400 =$
.02375 × 8400 = $199.50

APR 28

$\left(\sqrt[3]{17}\sqrt{3}\right)^2 = x^2 + 2x^2$
$3\left(17^{\frac{2}{3}}\right) = 3x^2$
$17^{\frac{1}{3}} = x$
vol. of cube $= x^3 = \left(17^{\frac{1}{3}}\right)^3 = 17$

APR 30 Find the minimum value by either finding the axis of symmetry or by finding where the first derivative is 0.

$y = 2x^2 - 4x + 25$
$y' = 4x - 4$
$4x - 4 = 0$
$x = 1$
find y for $x = 1$
$2 \cdot 1^2 - 4 \cdot 1 + 25 = 23$

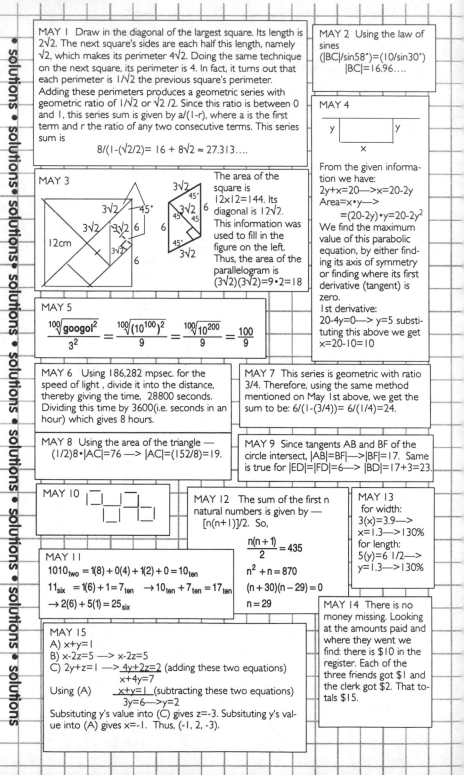

MAY 1 Draw in the diagonal of the largest square. Its length is $2\sqrt{2}$. The next square's sides are each half this length, namely $\sqrt{2}$, which makes its perimeter $4\sqrt{2}$. Doing the same technique on the next square, its perimeter is 4. In fact, it turns out that each perimeter is $1/\sqrt{2}$ the previous square's perimeter. Adding these perimeters produces a geometric series with geometric ratio of $1/\sqrt{2}$ or $\sqrt{2}/2$. Since this ratio is between 0 and 1, this series sum is given by $a/(1-r)$, where a is the first term and r the ratio of any two consecutive terms. This series sum is
$$8/(1-(\sqrt{2}/2)= 16 + 8\sqrt{2} \approx 27.313....$$

MAY 2 Using the law of sines
$(|BC|/\sin58°)=(10/\sin30°)$
$|BC|=16.96....$

MAY 4

y ⬜ y
x

From the given information we have:
$2y+x=20 \longrightarrow x=20-2y$
$Area=x\cdot y \longrightarrow$
$=(20-2y)\cdot y=20-2y^2$
We find the maximum value of this parabolic equation, by either finding its axis of symmetry or finding where its first derivative (tangent) is zero.
1st derivative:
$20-4y=0 \longrightarrow y=5$ substituting this above we get $x=20-10=10$

MAY 3

The area of the square is $12\times12=144$. Its diagonal is $12\sqrt{2}$. This information was used to fill in the figure on the left. Thus, the area of the parallelogram is $(3\sqrt{2})(3\sqrt{2})=9\cdot2=18$

MAY 5
$$\frac{\sqrt[100]{googol^2}}{3^2} = \frac{\sqrt[100]{(10^{100})^2}}{9} = \frac{\sqrt[100]{10^{200}}}{9} = \frac{100}{9}$$

MAY 6 Using 186,282 mpsec. for the speed of light , divide it into the distance, thereby giving the time, 28800 seconds. Dividing this time by 3600(i.e. seconds in an hour) which gives 8 hours.

MAY 7 This series is geometric with ratio 3/4. Therefore, using the same method mentioned on May 1st above, we get the sum to be: $6/(1-(3/4))= 6/(1/4)=24$.

MAY 8 Using the area of the triangle —
$(1/2)8\cdot|AC|=76 \longrightarrow |AC|=(152/8)=19$.

MAY 9 Since tangents AB and BF of the circle intersect, $|AB|=BF| \longrightarrow |BF|=17$. Same is true for $|ED|=|FD|=6 \longrightarrow |BD|=17+3=23$.

MAY 10

MAY 11
$1010_{two} = 1(8) + 0(4) + 1(2) + 0 = 10_{ten}$
$11_{six} = 1(6)+1=7_{ten} \rightarrow 10_{ten}+7_{ten}=17_{ten}$
$\rightarrow 2(6) + 5(1) = 25_{six}$

MAY 12 The sum of the first n natural numbers is given by —
$[n(n+1)]/2$. So,
$$\frac{n(n+1)}{2} = 435$$
$n^2 +n= 870$
$(n + 30)(n - 29) = 0$
$n = 29$

MAY 13
for width:
$3(x)=3.9 \longrightarrow$
$x=1.3 \longrightarrow 130\%$
for length:
$5(y)=6 1/2 \longrightarrow$
$y=1.3 \longrightarrow 130\%$

MAY 14 There is no money missing. Looking at the amounts paid and where they went we find: there is $10 in the register. Each of the three friends got $1 and the clerk got $2. That totals $15.

MAY 15
A) $x+y=1$
B) $x-2z=5 \longrightarrow x-2z=5$
C) $2y+z=1 \longrightarrow \underline{4y+2z=2}$ (adding these two equations)
$x+4y=7$
Using (A) $\underline{x+y=1}$ (subtracting these two equations)
$3y=6 \longrightarrow y=2$
Substituting y's value into (C) gives z=-3. Substituting y's value into (A) gives x=-1. Thus, (-1, 2, -3).

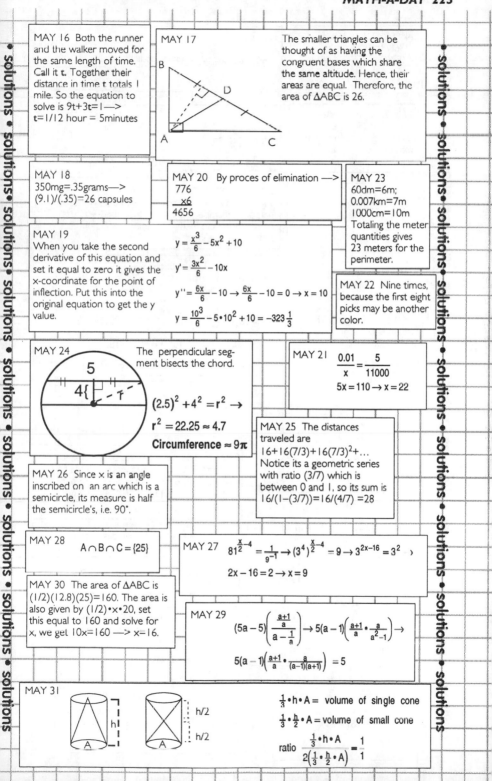

MAY 16 Both the runner and the walker moved for the same length of time. Call it **t**. Together their distance in time t totals 1 mile. So the equation to solve is $9t+3t=1 \longrightarrow$ $t=1/12$ hour = 5 minutes

MAY 17

The smaller triangles can be thought of as having the congruent bases which share the same altitude. Hence, their areas are equal. Therefore, the area of $\triangle ABC$ is 26.

MAY 18
350mg=.35grams $\longrightarrow$
$(9.1)/(.35)$=26 capsules

MAY 20 By proces of elimination $\longrightarrow$
$$\begin{array}{r} 776 \\ \times 6 \\ \hline 4656 \end{array}$$

MAY 23
60dm=6m;
0.007km=7m
1000cm=10m
Totaling the meter quantities gives 23 meters for the perimeter.

MAY 19
When you take the second derivative of this equation and set it equal to zero it gives the x-coordinate for the point of inflection. Put this into the original equation to get the y value.

$$y = \frac{x^3}{6} - 5x^2 + 10$$

$$y' = \frac{3x^2}{6} - 10x$$

$$y'' = \frac{6x}{6} - 10 \to \frac{6x}{6} - 10 = 0 \to x = 10$$

$$y = \frac{10^3}{6} - 5 \cdot 10^2 + 10 = -323\tfrac{1}{3}$$

MAY 22 Nine times, because the first eight picks may be another color.

MAY 24

5

4{

The perpendicular segment bisects the chord.

$$(2.5)^2 + 4^2 = r^2 \to$$

$$r^2 = 22.25 \approx 4.7$$

Circumference $\approx 9\pi$

MAY 21 $\dfrac{0.01}{x} = \dfrac{5}{11000}$

$5x = 110 \to x = 22$

MAY 25 The distances traveled are
$16+16(7/3)+16(7/3)^2+\ldots$
Notice its a geometric series with ratio $(3/7)$ which is between 0 and 1, so its sum is $16/(1-(3/7))=16/(4/7) =28$

MAY 26 Since x is an angle inscribed on an arc which is a semicircle, its measure is half the semicircle's, i.e. 90°.

MAY 28

$A \cap B \cap C = \{25\}$

MAY 27 $81^{\frac{x}{2}-4} = \frac{1}{9^{-1}} \to (3^4)^{\frac{x}{2}-4} = 9 \to 3^{2x-16} = 3^2$,

$2x - 16 = 2 \to x = 9$

MAY 30 The area of $\triangle ABC$ is $(1/2)(12.8)(25)=160$. The area is also given by $(1/2) \cdot x \cdot 20$, set this equal to 160 and solve for x, we get $10x=160 \longrightarrow x=16$.

MAY 29

$$(5a-5)\left(\frac{\frac{a+1}{a}}{a-\frac{1}{a}}\right) \to 5(a-1)\left(\frac{a+1}{a} \cdot \frac{a}{a^2-1}\right) \to$$

$$5(a-1)\left(\frac{a+1}{a} \cdot \frac{a}{(a-1)(a+1)}\right) = 5$$

MAY 31

h

h/2

h/2

$\frac{1}{3} \cdot h \cdot A$ = volume of single cone

$\frac{1}{3} \cdot \frac{h}{2} \cdot A$ = volume of small cone

ratio $\dfrac{\frac{1}{3} \cdot h \cdot A}{2\left(\frac{1}{3} \cdot \frac{h}{2} \cdot A\right)} = \dfrac{1}{1}$

JUN 1 (1472/4)=368
  (368/4)=92
  (92/4)=23 because the amount in the jar had quadrupled 3 times in 1 1/2 minute

JUN 2

$$10^? = \text{googolplex} = 10^{\text{googol}} = 10^{10^{100}}$$

JUN 3 26 minutes. After 25 minutes, it is 25 inches up the wall. During the 26th minute it climbs over the top and doesn't slide back.

JUN 4

$$\frac{9}{10} + \frac{9}{100} + \frac{9}{1000} + \dots$$
$$= 0.9 + 0.09 + 0.009 + \dots = 0.\overline{9} \quad \text{— whose limit is 1.}$$

JUN 6
CIX—>109
LXXXIV—>50+30+4=84
109-84=25

JUN 5 Lewis Carroll's solution is —
winter—>winner—>wanner—>warder—> harder—>
harper—>hamper—damper—>damped—>
dammed—> dimmed—>dimmer—>
simmer—> summer

JUN 7 The two triangle are similar because they have two pairs of corresponding angles congruent (the vertical angles and the given angles). Therefore, their corresponding sides are proportional.

$$\frac{x}{132} = \frac{5}{66}$$
$$x = 10$$

JUN 8 See Jan 17th solution for formula information.

$$A = P\left(1+\frac{r}{n}\right)^{nt}$$
$$A = 10\left(1+\frac{1}{n}\right)^{n \cdot 1}$$
$$= 10\left(1+\frac{1}{n}\right)^{n} \rightarrow 10 \cdot e \approx 10(2.7182\dots)$$
$$\text{as } n \rightarrow \infty \qquad \approx \$27.18$$

JUN 9 For this to be a real number the expression
39-3x≥0 —> x≤13

JUN 10

$$\left(37 + \frac{18}{n}\right) \cdot n = 721$$
$$37n + 18 = 721$$
$$n = 19$$

JUN 12 The circumference of the circle is $(6/\pi) \cdot \pi$=6". So (1/2)" per second means and angle of (1/2)/6 =1/12 of a revolution per second—> (1/12)(360°)=30° per second.

JUN 11

Writing everything in term of sin & cos, we get

$$7\left[2\frac{\frac{\sin\theta}{\cos\theta}}{\frac{1}{\cos\theta}}(2\sin\theta) + 4\frac{\cos\theta}{\frac{1}{\cos\theta}}\right] \rightarrow$$
$$7\left[4\sin^2\theta + 4\cos^2\theta\right] \rightarrow$$
$$7\left[4(\sin^2\theta + \cos^2\theta)\right] = 28$$

Note: $\sin^2\theta + \cos^2\theta = 1$

JUN 13 A nine point circle passes through the base of each altitude of the triangle and each midpoint of each of its sides. Its center is located at the intersection of the triangle's medians.

JUN 15 The diameter of the sphere will be the length of the cube's side. The sphere's surface area is

$$4\pi r^2 = 4\pi$$
$$r^2 = 1$$
$$r = 1 \rightarrow \text{diameter} = 2$$

Thus, each of the six faces of the cube has area 4. So the cubes surface area is
4•6=24

JUN 14

$$667 \div 23 = 29$$

JUN 16 "I came to be hanged." If the town judges him to be lying and hangs him, they will have hung a man who was telling the truth. —A paradox.

JUN 17 Each 2" tile is 4 sq.in.. So 10 tiles will do it if you are careful in cutting and placing the pieces in their proper space.

$$64 + x^2 = (2\sqrt{41})^2$$
$$64 + x^2 = 4 \cdot 41 \rightarrow x = 10$$
$$\text{area}\triangle = \frac{8 \cdot 10}{2} = 40 \text{ sq. in.}$$

JUN 18

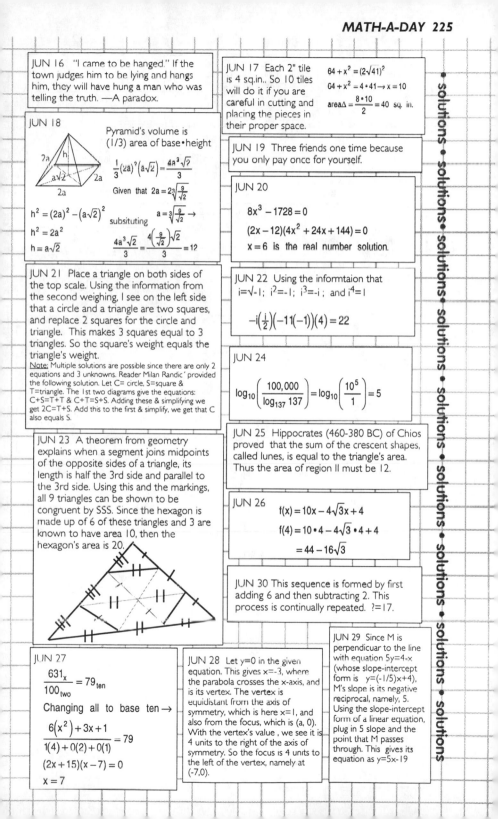

Pyramid's volume is (1/3) area of base • height

$$\frac{1}{3}(2a)^2 (a\sqrt{2}) = \frac{4a^3\sqrt{2}}{3}$$

Given that $2a = 2\sqrt[3]{\frac{9}{\sqrt{2}}}$

$$h^2 = (2a)^2 - (a\sqrt{2})^2 \qquad a = \sqrt[3]{\frac{9}{\sqrt{2}}} \rightarrow$$

substituting

$$h^2 = 2a^2$$

$$h = a\sqrt{2} \qquad \frac{4a^3\sqrt{2}}{3} = \frac{4\left(\frac{9}{\sqrt{2}}\right)\sqrt{2}}{3} = 12$$

JUN 19 Three friends one time because you only pay once for yourself.

JUN 20

$$8x^3 - 1728 = 0$$
$$(2x - 12)(4x^2 + 24x + 144) = 0$$
$$x = 6 \text{ is the real number solution.}$$

JUN 21 Place a triangle on both sides of the top scale. Using the information from the second weighing, I see on the left side that a circle and a triangle are two squares, and replace 2 squares for the circle and triangle. This makes 3 squares equal to 3 triangles. So the square's weight equals the triangle's weight.

Note: Multiple solutions are possible since there are only 2 equations and 3 unknowns. Reader Milan Randic' provided the following solution. Let C= circle, S=square & T=triangle. The 1st two diagrams give the equations: C+S=T+T & C+T=S+S. Adding these & simplifying we get 2C=T+S. Add this to the first & simplify, we get that C also equals S.

JUN 22 Using the informtaion that $i=\sqrt{-1}$; $i^2=-1$; $i^3=-i$; and $i^4=1$

$$-i\left(\frac{i}{2}\right)(-11(-1))(4) = 22$$

JUN 24

$$\log_{10}\left(\frac{100,000}{\log_{137} 137}\right) = \log_{10}\left(\frac{10^5}{1}\right) = 5$$

JUN 23 A theorem from geometry explains when a segment joins midpoints of the opposite sides of a triangle, its length is half the 3rd side and parallel to the 3rd side. Using this and the markings, all 9 triangles can be shown to be congruent by SSS. Since the hexagon is made up of 6 of these triangles and 3 are known to have area 10, then the hexagon's area is 20.

JUN 25 Hippocrates (460-380 BC) of Chios proved that the sum of the crescent shapes, called lunes, is equal to the triangle's area. Thus the area of region II must be 12.

JUN 26

$$f(x) = 10x - 4\sqrt{3}x + 4$$
$$f(4) = 10 \cdot 4 - 4\sqrt{3} \cdot 4 + 4$$
$$= 44 - 16\sqrt{3}$$

JUN 30 This sequence is formed by first adding 6 and then subtracting 2. This process is continually repeated. ?=17.

JUN 29 Since M is perpendicuar to the line with equation 5y=4-x (whose slope-intercept form is y=(-1/5)x+4), M's slope is its negative reciprocal, namely, 5. Using the slope-intercept form of a linear equation, plug in 5 slope and the point that M passes through. This gives its equation as y=5x-19

JUN 27

$$\frac{631_x}{100_{two}} = 79_{ten}$$

Changing all to base ten →

$$\frac{6(x^2) + 3x + 1}{1(4) + 0(2) + 0(1)} = 79$$

$$(2x + 15)(x - 7) = 0$$

$$x = 7$$

JUN 28 Let y=0 in the given equation. This gives x=-3, where the parabola crosses the x-axis, and is its vertex. The vertex is equidistant from the axis of symmetry, which is here x=1, and also from the focus, which is (a, 0). With the vertex's value, we see it is 4 units to the right of the axis of symmetry. So the focus is 4 unts to the left of the vertex, namely at (-7,0).

July 1
The objects are moving counter-clockwise with one object sides increasing with each move.

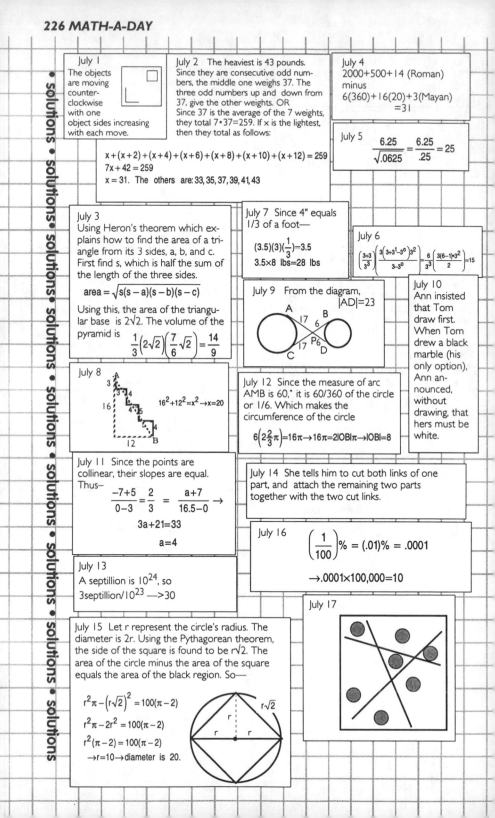

July 2 The heaviest is 43 pounds. Since they are consecutive odd numbers, the middle one weighs 37. The three odd numbers up and down from 37, give the other weights. Since 37 is the average of the 7 weights, they total $7 \cdot 37 = 259$. If x is the lightest, then they total as follows:

$$x + (x+2) + (x+4) + (x+6) + (x+8) + (x+10) + (x+12) = 259$$
$$7x + 42 = 259$$
$$x = 31. \text{ The others are: } 33, 35, 37, 39, 41, 43$$

July 4
2000+500+14 (Roman)
minus
6(360)+16(20)+3(Mayan)
=31

July 5
$$\frac{6.25}{\sqrt{.0625}} = \frac{6.25}{.25} = 25$$

July 3
Using Heron's theorem which explains how to find the area of a triangle from its 3 sides, a, b, and c. First find s, which is half the sum of the length of the three sides.

$$\text{area} = \sqrt{s(s-a)(s-b)(s-c)}$$

Using this, the area of the triangular base is $2\sqrt{2}$. The volume of the pyramid is $\frac{1}{3}\left(2\sqrt{2}\right)\left(\frac{7}{6}\sqrt{2}\right) = \frac{14}{9}$

July 7 Since 4" equals 1/3 of a foot—
$$(3.5)(3)\left(\frac{1}{3}\right)=3.5$$
$$3.5\times 8 \text{ lbs} = 28 \text{ lbs}$$

July 6
$$\left(\frac{3+3}{3^3}\right)\frac{3\left(3+3^1-3^0\right)3^2}{3-3^0} = \frac{6}{3^3}\left(\frac{3(6-1)\times 3^2}{2}\right)=15$$

July 9 From the diagram, |AD|=23

July 10
Ann insisted that Tom draw first. When Tom drew a black marble (his only option), Ann announced, without drawing, that hers must be white.

July 8
$$16^2+12^2=x^2 \rightarrow x=20$$

July 12 Since the measure of arc AMB is 60,° it is 60/360 of the circle or 1/6. Which makes the circumference of the circle
$$6\left(2\frac{2}{3}\pi\right)=16\pi \rightarrow 16\pi=2|OB|\pi \rightarrow |OB|=8$$

July 11 Since the points are collinear, their slopes are equal. Thus—
$$\frac{-7+5}{0-3} = \frac{2}{3} = \frac{a+7}{16.5-0} \rightarrow$$
$$3a+21=33$$
$$a=4$$

July 14 She tells him to cut both links of one part, and attach the remaining two parts together with the two cut links.

July 16
$$\left(\frac{1}{100}\right)\% = (.01)\% = .0001$$
$$\rightarrow .0001 \times 100,000 = 10$$

July 13
A septillion is 10^{24}, so
3septillion/10^{23} —>30

July 17

July 15 Let r represent the circle's radius. The diameter is 2r. Using the Pythagorean theorem, the side of the square is found to be $r\sqrt{2}$. The area of the circle minus the area of the square equals the area of the black region. So—

$$r^2\pi - \left(r\sqrt{2}\right)^2 = 100(\pi - 2)$$
$$r^2\pi - 2r^2 = 100(\pi - 2)$$
$$r^2(\pi - 2) = 100(\pi - 2)$$
$$\rightarrow r=10 \rightarrow \text{diameter is } 20.$$

July 18 $y = 9$

July 19 Using the law of cosines—

$$21^2 = 18^2 + 20^2 - 2(18)(20)\cos a$$

$$-283 = -720\cos a$$

$$.393055\ldots = \cos a$$

$$67° \approx 66.85\ldots° = a$$

July 21 Since x is a central angle, its intercepted arc is $x°$. The external angle of a circle equals half the difference of its intercepted arcs, namely

$$\frac{x-11°}{2} = 9° \rightarrow x = 29°$$

July 20

$$\left(\sqrt[3]{8x^6}\sqrt{2}\right)^{256} = \frac{1}{\frac{4}{4}x^{-\frac{1}{2}}} \rightarrow \left(2x^2\sqrt{2}\right)^{256} - \frac{1}{\left(2^2\right)^{\left(\frac{8-x}{2x}\right)}}$$

$$\left(2^{\frac{1}{2x^2}}\right)^{256} = (2)^{\frac{-8+x}{x}} \rightarrow$$

for x>0 we get

$$\frac{128}{x^2} = \frac{x-8}{x} \rightarrow 128x = x^3 - 8x^2 \rightarrow 0 = x\left(x^2 - 8x - 128\right) \rightarrow x = 16$$

July 22

$$\frac{\frac{4x-\frac{1}{x}}{1+\frac{2}{2x+1}}}{\frac{4x^2+4x+1}{3x}} = \frac{\frac{\frac{4x^2-1}{x}}{\frac{2x+1-2}{2x+1}}}{\frac{(2x+1)(2x+1)}{3x}} = \frac{\frac{(2x-1)(2x+1)}{x}}{\frac{2x-1}{2x+1}} = \frac{(2x+1)(2x+1)}{3x}$$

$$\frac{(2x-1)(2x+1)}{x} \cdot \frac{(2x+1)}{(2x-1)} \cdot \frac{3x}{(2x+1)(2x+1)} = 3$$

July 24 It is Saturday. No other day works with the statement "I told the truth yesterday".

July 23 On an incline, gravity is resolved by a force parallel to the incline of the plane and one perpendicular to the incline of the plane.

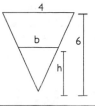

Force of gravity is resolved into two vectors. The gravity force is the weight of the truck—

force of brakes: x

force of gravity 3400

Using triangle ABC—>

$$\sin 30° = x/3400$$

$$x = 1700$$

July 25 Complete the square, and rearrange the equation into the following form—

$$4(x-3)^2 + (y+1)^2 = 16$$

$$\frac{(x-3)^2}{4} + \frac{(y+1)^2}{16} = 1 \rightarrow$$

$$\sqrt{16} = 4$$

the major axis is twice this or 8

July 26 Since time=distance/speed, the slow car will have traveled (108/55) hours ≈117.8 minutes to go 108 miles at a speed of 55mph. The fast car will have traveled (108/65)hours ≈ 99.7 minutes after 108 miles. The difference of these is a little more than 18 minutes.

July 27 It is 27. Its cube is 19683, and these digits total 27.

July 29   Note: the height of the sphere equals its diameter.

$$V_{cone} = V_{sphere}$$

$$\pi r^2 \left(22\sqrt{6}\right) = \frac{4}{3}\pi\left(11\sqrt{6}\right)^3$$

$$r^2 = 484$$

$$r = 22$$

July 30

July 28 Using the longest side, and dividing into 4 equal parts gives the greater volume—

$$(2.5)^2(4.96) = 31$$

July 31
Let h represent the height of the water.

Note: the volume of the water is equal to the prism's volume with triangular base and altitude the length 10 of the prism.

similar triangles give

$$\frac{4}{6} = \frac{b}{h} \rightarrow b = \frac{2}{3}h$$

$$\text{Volume}_{water} = \left(\frac{1}{2}h \cdot b\right) \cdot \text{altitude} = \left(\frac{1}{2}h \cdot \frac{2}{3}h\right) \cdot 10$$

$$83\frac{1}{3} = \frac{2}{6}h^2 \cdot 10 \rightarrow h = 5, \text{ the height of the water}$$

AUG 1 Either line up the rulers and notice it looks like it approaches 2. Or look at the rulers as the series— $1+(1/2)+(1/4)+(1/8)...$ which is geometric with ratio $(1/2)$ so its sum is $1/(1-(1/2)) =2$

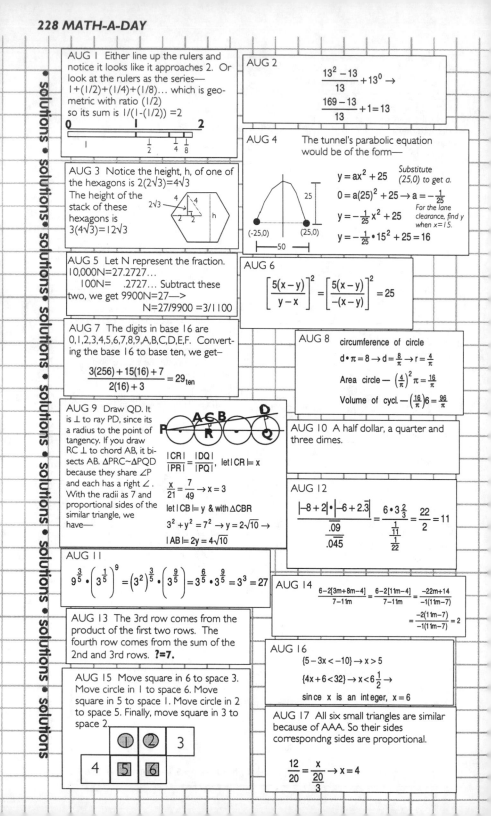

AUG 2

$$\frac{13^2 - 13}{13} + 13^0 \rightarrow$$

$$\frac{169 - 13}{13} + 1 = 13$$

AUG 3 Notice the height, h, of one of the hexagons is $2(2\sqrt{3})=4\sqrt{3}$
The height of the stack of these hexagons is $3(4\sqrt{3})=12\sqrt{3}$

AUG 4 The tunnel's parabolic equation would be of the form—

$$y = ax^2 + 25$$

Substitute $(25,0)$ to get a.

$$0 = a(25)^2 + 25 \rightarrow a = -\frac{1}{25}$$

For the lane clearance, find y when x=15.

$$y = -\frac{1}{25}x^2 + 25$$

$$y = -\frac{1}{25} \cdot 15^2 + 25 = 16$$

AUG 5 Let N represent the fraction.
$10,000N=27.2727...$
$100N= .2727...$ Subtract these two, we get $9900N=27\rightarrow$
$N=27/9900 =3/1100$

AUG 6

$$\left[\frac{5(x-y)}{y-x}\right]^2 = \left[\frac{5(x-y)}{-(x-y)}\right]^2 = 25$$

AUG 7 The digits in base 16 are $0,1,2,3,4,5,6,7,8,9,A,B,C,D,E,F$. Converting the base 16 to base ten, we get—

$$\frac{3(256) + 15(16) + 7}{2(16) + 3} = 29_{ten}$$

AUG 8 circumference of circle
$$d \cdot \pi = 8 \rightarrow d = \frac{8}{\pi} \rightarrow r = \frac{4}{\pi}$$
Area circle — $\left(\frac{4}{\pi}\right)^2 \pi = \frac{16}{\pi}$
Volume of cycl. — $\left(\frac{16}{\pi}\right)6 = \frac{96}{\pi}$

AUG 9 Draw QD. It is ⊥ to ray PD, since its a radius to the point of tangency. If you draw RC ⊥ to chord AB, it bisects AB. $\triangle PRC \sim \triangle PQD$ because they share $\angle P$ and each has a right $\angle$. With the radii as 7 and proportional sides of the similar triangle, we have—

$$\frac{|CR|}{|PR|} = \frac{|DQ|}{|PQ|}, \text{ let } |CR| = x$$

$$\frac{x}{21} = \frac{7}{49} \rightarrow x = 3$$

let $|CB| = y$ & with $\triangle CBR$

$$3^2 + y^2 = 7^2 \rightarrow y = 2\sqrt{10} \rightarrow$$

$$|AB| = 2y = 4\sqrt{10}$$

AUG 10 A half dollar, a quarter and three dimes.

AUG 12

$$\frac{|-8+2| \cdot |-6+2.\overline{3}|}{\frac{.09}{.045}} = \frac{6 \cdot 3\frac{2}{3}}{\frac{1}{11}} = \frac{22}{2} = 11$$

AUG 11

$$9^{\frac{3}{5}} \cdot \left(3^{\frac{1}{5}}\right)^9 = \left(3^2\right)^{\frac{3}{5}} \cdot \left(3^{\frac{9}{5}}\right) = 3^{\frac{6}{5}} \cdot 3^{\frac{9}{5}} = 3^3 = 27$$

AUG 14

$$\frac{6-2[3m+8m-4]}{7-11m} = \frac{6-2[11m-4]}{7-11m} = \frac{-22m+14}{-1(11m-7)}$$

$$= \frac{-2(11m-7)}{-1(11m-7)} = 2$$

AUG 13 The 3rd row comes from the product of the first two rows. The fourth row comes from the sum of the 2nd and 3rd rows. **?=7.**

AUG 16
$$\{5 - 3x < -10\} \rightarrow x > 5$$
$$\{4x + 6 < 32\} \rightarrow x < 6\frac{1}{2} \rightarrow$$
since x is an integer, $x = 6$

AUG 15 Move square in 6 to space 3. Move circle in 1 to space 6. Move square in 5 to space 1. Move circle in 2 to space 5. Finally, move square in 3 to space 2.

AUG 17 All six small triangles are similar because of AAA. So their sides correspondng sides are proportional.

$$\frac{12}{20} = \frac{x}{\frac{20}{3}} \rightarrow x = 4$$

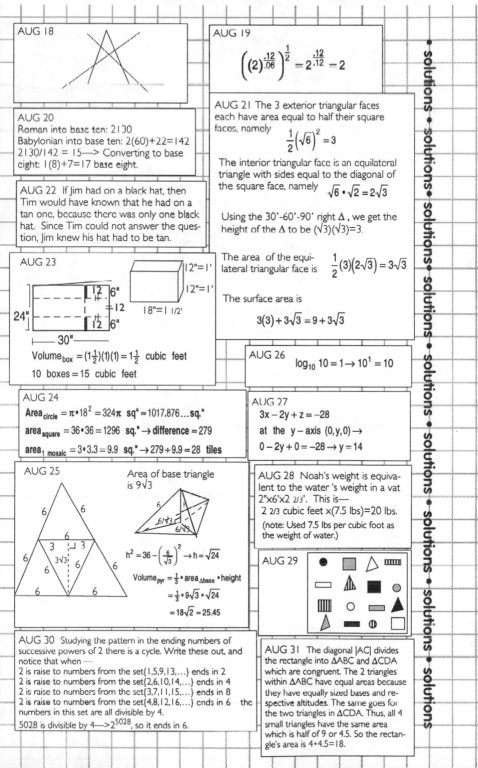

AUG 18

AUG 19

$$\left((2)^{\frac{.12}{.06}}\right)^{\frac{1}{2}} = 2^{\frac{.12}{.12}} = 2$$

AUG 20
Roman into base ten: 2130
Babylonian into base ten: 2(60)+22=142
2130/142 = 15—> Converting to base
eight: 1(8)+7=17 base eight.

AUG 21 The 3 exterior triangular faces each have area equal to half their square faces, namely

$$\frac{1}{2}\left(\sqrt{6}\right)^2 = 3$$

The interior triangular face is an equilateral triangle with sides equal to the diagonal of the square face, namely $\sqrt{6} \cdot \sqrt{2} = 2\sqrt{3}$

Using the 30°-60°-90° right Δ, we get the height of the Δ to be $(\sqrt{3})(\sqrt{3})=3$.

AUG 22 If Jim had on a black hat, then Tim would have known that he had on a tan one, because there was only one black hat. Since Tim could not answer the question, Jim knew his hat had to be tan.

The area of the equilateral triangular face is $\frac{1}{2}(3)(2\sqrt{3}) = 3\sqrt{3}$

The surface area is

$$3(3)+3\sqrt{3} = 9 + 3\sqrt{3}$$

AUG 23

12"=1'
12"=1'
18"=1 1/2'
24"
12 6"
12 6"
12 =12
30"

Volume$_{box}$ = $(1\frac{1}{2})(1)(1) = 1\frac{1}{2}$ cubic feet
10 boxes = 15 cubic feet

AUG 26

$$\log_{10} 10 = 1 \rightarrow 10^1 = 10$$

AUG 24

Area$_{circle}$ = $\pi \cdot 18^2 = 324\pi$ sq." $\approx 1017.876...$ sq."
area$_{square}$ = $36 \cdot 36 = 1296$ sq." $\rightarrow$ difference ≈ 279
area$_{1\ mosaic}$ = $3 \cdot 3.3 = 9.9$ sq." $\rightarrow 279 \div 9.9 \approx 28$ tiles

AUG 27
$3x - 2y + z = -28$
at the y-axis $(0,y,0) \rightarrow$
$0 - 2y + 0 = -28 \rightarrow y = 14$

AUG 25

Area of base triangle is $9\sqrt{3}$

6
6
6
3 3
6 3√3 6
6
6 6

$h^2 = 36 - \left(\frac{6}{\sqrt{3}}\right)^2 \rightarrow h = \sqrt{24}$

Volume$_{pyr}$ = $\frac{1}{3} \cdot$ area$_{\Delta base} \cdot$ height
= $\frac{1}{3} \cdot 9\sqrt{3} \cdot \sqrt{24}$
= $18\sqrt{2} \approx 25.45$

AUG 28 Noah's weight is equivalent to the water's weight in a vat 2"x6'x2 2/3'. This is—
2 2/3 cubic feet x(7.5 lbs)=20 lbs.
(note: Used 7.5 lbs per cubic foot as the weight of water.)

AUG 29

AUG 30 Studying the pattern in the ending numbers of successive powers of 2 there is a cycle. Write these out, and notice that when —
2 is raise to numbers from the set{1,5,9,13,...} ends in 2
2 is raise to numbers from the set{2,6,10,14,...} ends in 4
2 is raise to numbers from the set{3,7,11,15,...} ends in 8
2 is raise to numbers from the set{4,8,12,16,...} ends in 6 the numbers in this set are all divisible by 4.
5028 is divisible by 4—>2^{5028}, so it ends in 6.

AUG 31 The diagonal |AC| divides the rectangle into ΔABC and ΔCDA which are congruent. The 2 triangles within ΔABC have equal areas because they have equally sized bases and respective altitudes. The same goes for the two triangles in ΔCDA. Thus, all 4 small triangles have the same area which is half of 9 or 4.5. So the rectangle's area is $4 \cdot 4.5 = 18$.

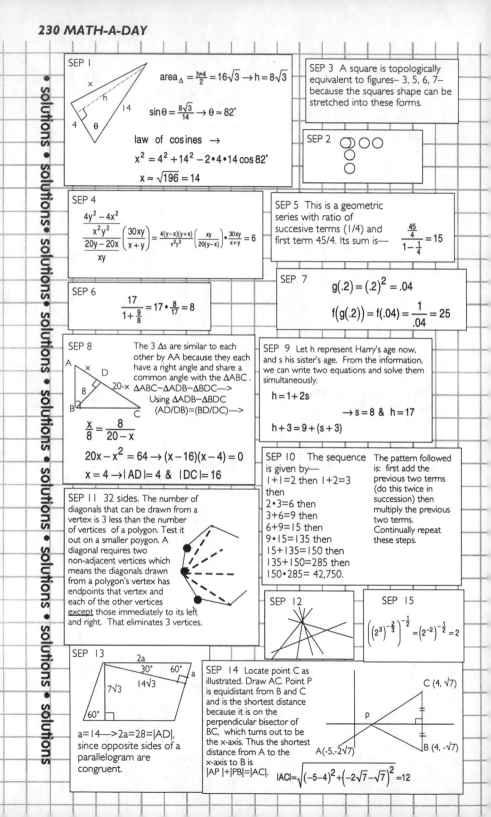

SEP 1

$$\text{area}_\Delta = \frac{h \cdot 4}{2} = 16\sqrt{3} \rightarrow h = 8\sqrt{3}$$

$$\sin\theta = \frac{8\sqrt{3}}{14} \rightarrow \theta \approx 82°$$

law of cosines →

$$x^2 = 4^2 + 14^2 - 2 \cdot 4 \cdot 14 \cos 82°$$

$$x \approx \sqrt{196} = 14$$

SEP 3 A square is topologically equivalent to figures– 3, 5, 6, 7– because the squares shape can be stretched into these forms.

SEP 2

SEP 4

$$\frac{\dfrac{4y^2 - 4x^2}{x^2 y^2}}{\dfrac{20y - 20x}{xy}} \left(\frac{30xy}{x+y} \right) = \frac{4(y-x)(y+x)}{x^2 y^2} \left(\frac{xy}{20(y-x)} \right) \cdot \frac{30xy}{x+y} = 6$$

SEP 5 This is a geometric series with ratio of succesive terms (1/4) and first term 45/4. Its sum is—

$$\frac{\frac{45}{4}}{1 - \frac{1}{4}} = 15$$

SEP 6

$$\frac{17}{1 + \frac{9}{8}} = 17 \cdot \frac{8}{17} = 8$$

SEP 7

$$g(.2) = (.2)^2 = .04$$

$$f\big(g(.2)\big) = f(.04) = \frac{1}{.04} = 25$$

SEP 8

The 3 Δs are similar to each other by AA because they each have a right angle and share a common angle with the ΔABC .
ΔABC~ΔADB~ΔBDC—>
Using ΔADB~ΔBDC
(AD/DB)=(BD/DC)—>

$$\frac{x}{8} = \frac{8}{20 - x}$$

$$20x - x^2 = 64 \rightarrow (x - 16)(x - 4) = 0$$

$$x = 4 \rightarrow |AD| = 4 \ \& \ |DC| = 16$$

SEP 9 Let h represent Harry's age now, and s his sister's age. From the information, we can write two equations and solve them simultaneously.

$$h = 1 + 2s$$

$$h + 3 = 9 + (s + 3)$$

→ s = 8 & h = 17

SEP 10   The sequence is given by—
1+1=2 then 1+2=3 then
2·3=6 then
3+6=9 then
6+9=15 then
9·15=135 then
15+135=150 then
135+150=285 then
150·285= 42,750.

The pattern followed is: first add the previous two terms (do this twice in succession) then multiply the previous two terms. Continually repeat these steps.

SEP 11   32 sides. The number of diagonals that can be drawn from a vertex is 3 less than the number of vertices of a polygon. Test it out on a smaller poygon. A diagonal requires two non-adjacent vertices which means the diagonals drawn from a polygon's vertex has endpoints that vertex and each of the other vertices except those immediately to its left and right. That eliminates 3 vertices.

SEP 12

SEP 15

$$\left(\left(2^3 \right)^{-\frac{2}{3}} \right)^{-\frac{1}{2}} = \left(2^{-2} \right)^{-\frac{1}{2}} = 2$$

SEP 13

a=14—>2a=28=|AD|, since opposite sides of a parallelogram are congruent.

SEP 14 Locate point C as illustrated. Draw AC. Point P is equidistant from B and C and is the shortest distance because it is on the perpendicular bisector of BC, which turns out to be the x-axis. Thus the shortest distance from A to the x-axis to B is |AP|+|PB|=|AC|.

C (4, √7)

P

A(-5,-2√7)   B (4, -√7)

$$|AC| = \sqrt{(-5-4)^2 + \left(-2\sqrt{7} - \sqrt{7}\right)^2} = 12$$

SEP 16 This is a famous paradox. There is no answer. If the barber shaves himself, he is shaving someone outside the group of "those who don't shave themselves." If he doesn't shave himself, he is a member of that group and therefore should shave himself.

SEP 18 These letters from the English alphabet all have a vertical line of symmetry. The two missing letters are M and T.

SEP 20

$|AB|=?$

$|AC|=10$

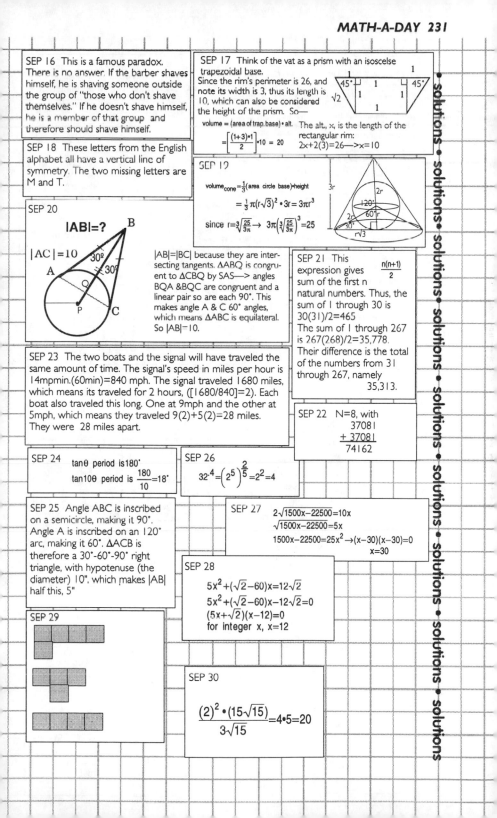

$|AB|=|BC|$ because they are intersecting tangents. $\triangle ABQ$ is congruent to $\triangle CBQ$ by SAS—> angles BQA &BQC are congruent and a linear pair so are each 90°. This makes angle A & C 60° angles, which means $\triangle ABC$ is equilateral. So $|AB|=10$.

SEP 17 Think of the vat as a prism with an isoscelse trapezoidal base. Since the rim's perimeter is 26, and note its width is 3, thus its length is 10, which can also be considered the height of the prism. So—

volume = (area of trap. base) • alt.

$$=\left[\frac{(1+3)\cdot 1}{2}\right]\cdot 10 = 20$$

The alt, x, is the length of the rectangular rim:

$2x+2(3)=26\rightarrow x=10$

SEP 19

$\text{volume}_{cone}=\frac{1}{3}(\text{area circle base})\cdot\text{height}$

$$=\frac{1}{3}\pi(r\sqrt{3})^2\cdot 3r=3\pi r^3$$

since $r=\sqrt[3]{\frac{25}{3\pi}}\rightarrow 3\pi\left(\sqrt[3]{\frac{25}{3\pi}}\right)^3=25$

SEP 21 This expression gives $\frac{n(n+1)}{2}$ sum of the first n natural numbers. Thus, the sum of 1 through 30 is 30(31)/2=465. The sum of 1 through 267 is 267(268)/2=35,778. Their difference is the total of the numbers from 31 through 267, namely 35,313.

SEP 22 N=8, with

```
   37081
 + 37081
   74162
```

SEP 23 The two boats and the signal will have traveled the same amount of time. The signal's speed in miles per hour is 14mpmin.(60min)=840 mph. The signal traveled 1680 miles, which means its traveled for 2 hours, ([1680/840]=2). Each boat also traveled this long. One at 9mph and the other at 5mph, which means they traveled 9(2)+5(2)=28 miles. They were 28 miles apart.

SEP 24 $\tan\theta$ period is180°

$\tan 10\theta$ period is $\frac{180}{10}=18°$

SEP 26

$32^{-4}=\left(2^5\right)^{\frac{2}{5}}=2^2=4$

SEP 25 Angle ABC is inscribed on a semicircle, making it 90°. Angle A is inscribed on an 120° arc, making it 60°. $\triangle ACB$ is therefore a 30°-60°-90° right triangle, with hypotenuse (the diameter) 10", which makes $|AB|$ half this, 5"

SEP 27

$2\sqrt{1500x-22500}=10x$

$\sqrt{1500x-22500}=5x$

$1500x-22500=25x^2\rightarrow(x-30)(x-30)=0$

$x=30$

SEP 28

$5x^2+(\sqrt{2}-60)x=12\sqrt{2}$

$5x^2+(\sqrt{2}-60)x-12\sqrt{2}=0$

$(5x+\sqrt{2})(x-12)=0$

for integer x, x=12

SEP 29

SEP 30

$$\frac{(2)^2\cdot(15\sqrt{15})}{3\sqrt{15}}=4\cdot 5=20$$

OCT 1 Tim is the rollerblader, Tom is the car driver, and Ted is the biker. WHY?—
We know Tom is not the rollerblader or the youngest. Ted is not the car driver. Ted is not the rollerblader because he is not the youngest. So that leaves Tim as the rollerblader, Tom the car driver, and Ted the biker.

OCT 2 Find the height of the equilateral triangle, and add 10cm for the radius above and below it.
$$=10\sqrt{3}+10$$

OCT 3
$$|BC|^2=\left(2\sqrt{29}\right)^2-4^2 \rightarrow |BC|=10$$
$$\textbf{area } \Delta ABC =\tfrac{1}{2}\cdot 4 \cdot 10 = 20$$
The area of ΔBCD is 10— half that of ΔABC because ΔABD & ΔBCD have the same size base and altitude, making their areas equal.

OCT 4 The tree and the stick form similar triangles, so their corresponding sides are congruent.
$$\frac{\text{tree ht.}}{36\tfrac{2}{3}}=\frac{3}{1\tfrac{2}{3}}\rightarrow \text{tree ht.}=66$$

OCT 5 The shortest side of a triangle is opposite its smallest angle. In ΔABE that is side BE. In ΔBCE, it is CE. In ΔCDE it is ED.
—> ED is the shortest.

OCT 6 Draw in a radius from the center to where x touches the circle. —>
$$3^2+x^2=\left(3\sqrt{10}\right)^2$$
$$x^2=81\rightarrow x=9$$

OCT 7 Labeling the vertices as shown. There are 12 routes.

ABCDE   ABMDE
ABMFE   ABCDMFE
ABCDMHGFE
ABMHGFE
plus the mirror images of these which start with AH

OCT 8
$$(3!)(\sqrt{9})(0.\bar{3})=6\cdot 3\cdot \tfrac{1}{3}=6$$

OCT 9
$$\left(\left(\left(64^{\frac{1}{2}}\right)^{\frac{1}{3}}\right)^{\frac{1}{4}}\right)^8=64^{\frac{8}{24}}=64^{\frac{1}{3}}=4$$

OCT 10
started with 1.0 g
6800mg= 6.8 g
120cg = 1.2 g
0.008kg = 8.0 g
total =17.0 g

OCT 11
Pictured is a flattened dodecahedron,. The diagram is called a Schlegel diagram. The 12 pentagonal faces of the dodecahedron are there, along with its 20 vertices. If each vertex is truncated., the resulting solid has 20+12
=32 faces.

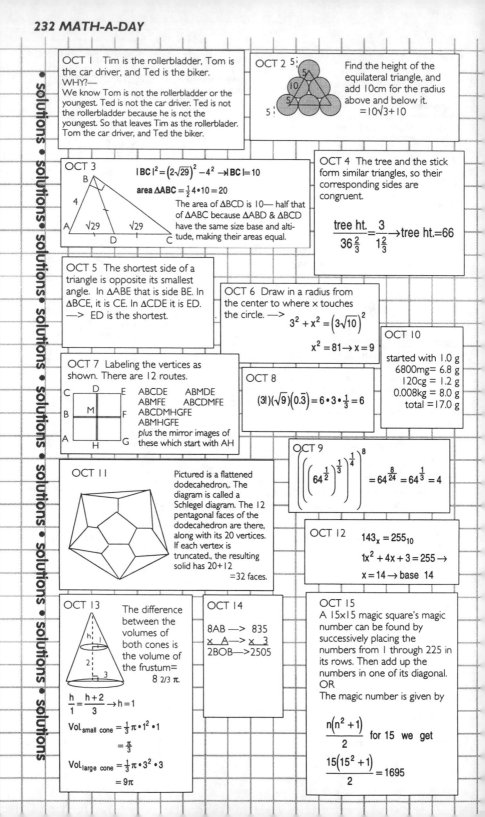

OCT 12
$$143_x=255_{10}$$
$$1x^2+4x+3=255\rightarrow$$
$$x=14\rightarrow \text{base } 14$$

OCT 13
The difference between the volumes of both cones is the volume of the frustum=
$8\,2/3\,\pi.$
$$\frac{h}{1}=\frac{h+2}{3}\rightarrow h=1$$
$$\text{Vol.}_{\text{small cone}}=\tfrac{1}{3}\pi\cdot 1^2\cdot 1$$
$$=\tfrac{\pi}{3}$$
$$\text{Vol.}_{\text{large cone}}=\tfrac{1}{3}\pi\cdot 3^2\cdot 3$$
$$=9\pi$$

OCT 14
$$\begin{array}{r} 8AB \rightarrow 835 \\ \underline{\times\ \ A} \rightarrow \underline{\times\ 3} \\ 2BOB \rightarrow 2505 \end{array}$$

OCT 15
A 15x15 magic square's magic number can be found by successively placing the numbers from 1 through 225 in its rows. Then add up the numbers in one of its diagonal.
OR
The magic number is given by
$$\frac{n(n^2+1)}{2}\quad \text{for 15 we get}$$
$$\frac{15(15^2+1)}{2}=1695$$

OCT 16

$$\left(6^2\right)^{2x-8} = 6^{3x} \rightarrow 4x - 16 = 3x$$
$$x = 16$$

OCT 17

$$3 = 9\left(\frac{5}{a+1}\right)$$
$$3 = \frac{45}{a+1}$$
$$3a + 3 = 45$$
$$a = 14$$

OCT 19

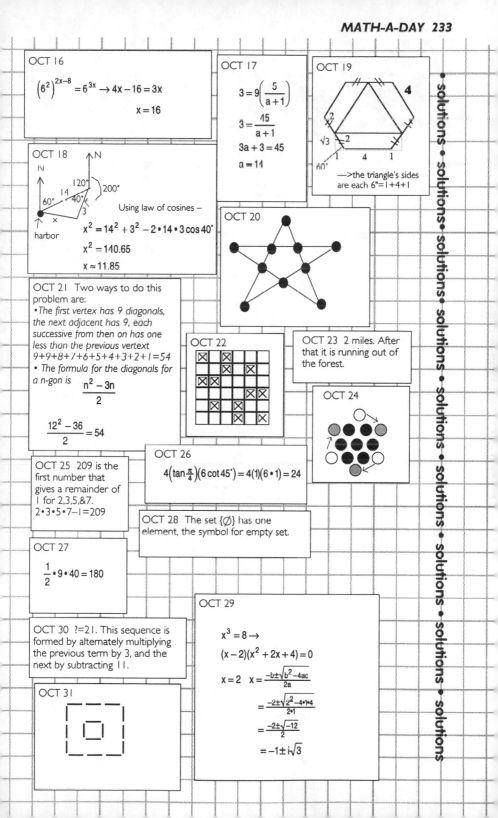

—>the triangle's sides
are each 6"=1+4+1

OCT 18

Using law of cosines –
$$x^2 = 14^2 + 3^2 - 2 \cdot 14 \cdot 3 \cos 40°$$
$$x^2 = 140.65$$
$$x \approx 11.85$$

OCT 20

OCT 21 Two ways to do this problem are:
• The first vertex has 9 diagonals, the next adjacent has 9, each successive from then on has one less than the previous vertext
9+9+8+7+6+5+4+3+2+1=54
• The formula for the diagonals for a n-gon is $\dfrac{n^2 - 3n}{2}$

$$\frac{12^2 - 36}{2} = 54$$

OCT 22

OCT 23 2 miles. After that it is running out of the forest.

OCT 24

OCT 25 209 is the first number that gives a remainder of 1 for 2,3,5,&7.
2•3•5•7−1=209

OCT 26

$$4\left(\tan\tfrac{\pi}{4}\right)\left(6\cot 45°\right) = 4(1)(6 \cdot 1) = 24$$

OCT 28 The set {∅} has one element, the symbol for empty set.

OCT 27

$$\frac{1}{2} \cdot 9 \cdot 40 = 180$$

OCT 30 ?=21. This sequence is formed by alternately multiplying the previous term by 3, and the next by subtracting 11.

OCT 29

$$x^3 = 8 \rightarrow$$
$$(x-2)(x^2 + 2x + 4) = 0$$
$$x = 2 \quad x = \frac{-b \pm \sqrt{b^2 - 4ac}}{2a}$$
$$= \frac{-2 \pm \sqrt{2^2 - 4 \cdot 1 \cdot 4}}{2 \cdot 1}$$
$$= \frac{-2 \pm \sqrt{-12}}{2}$$
$$= -1 \pm i\sqrt{3}$$

OCT 31

NOV 1 Put 3 coins on each side of the scale. The lighter stack has the lighter coin. Then weigh 2 of these 3 coins. If these two balance, the unweighed coin is the lighter one. Otherwise the lighter one is on the light side of the scale.

NOV 2 15min=1/4 of an hour. The distance the faster train travels in 1/4 hr. plus the distance the slower one does would give the distance between the two trains.—> (1/4)48 +(1/4)40 = 22 miles.

NOV 3

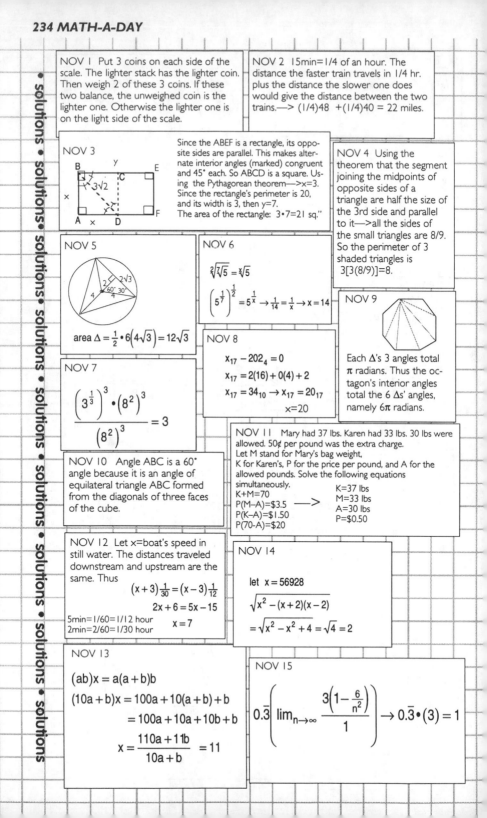

Since the ABEF is a rectangle, its opposite sides are parallel. This makes alternate interior angles (marked) congruent and 45° each. So ABCD is a square. Using the Pythagorean theorem—>x=3. Since the rectangle's perimeter is 20, and its width is 3, then y=7. The area of the rectangle: 3•7=21 sq."

NOV 4 Using the theorem that the segment joining the midpoints of opposite sides of a triangle are half the size of the 3rd side and parallel to it—>all the sides of the small triangles are 8/9. So the perimeter of 3 shaded triangles is 3[3(8/9)]=8.

NOV 5

area $\Delta = \frac{1}{2} \cdot 6\left(4\sqrt{3}\right) = 12\sqrt{3}$

NOV 6

$\sqrt[2]{\sqrt[7]{5}} = \sqrt[x]{5}$

$\left(5^{\frac{1}{7}}\right)^{\frac{1}{2}} = 5^{\frac{1}{x}} \rightarrow \frac{1}{14} = \frac{1}{x} \rightarrow x = 14$

NOV 9

Each Δ's 3 angles total π radians. Thus the octagon's interior angles total the 6 Δs' angles, namely 6π radians.

NOV 8

$x_{17} - 202_4 = 0$

$x_{17} = 2(16) + 0(4) + 2$

$x_{17} = 34_{10} \rightarrow x_{17} = 20_{17}$

$x = 20$

NOV 7

$$\frac{\left(3^{\frac{1}{3}}\right)^3 \cdot \left(8^2\right)^3}{\left(8^2\right)^3} = 3$$

NOV 10 Angle ABC is a 60° angle because it is an angle of equilateral triangle ABC formed from the diagonals of three faces of the cube.

NOV 11 Mary had 37 lbs. Karen had 33 lbs. 30 lbs were allowed. 50¢ per pound was the extra charge. Let M stand for Mary's bag weight, K for Karen's, P for the price per pound, and A for the allowed pounds. Solve the following equations simultaneously.

K=70
P(M–A)=$3.5
P(K–A)=$1.50
P(70-A)=$20

———>

K=37 lbs
M=33 lbs
A=30 lbs
P=$0.50

NOV 12 Let x=boat's speed in still water. The distances traveled downstream and upstream are the same. Thus

$(x+3)\frac{1}{30} = (x-3)\frac{1}{12}$

$2x + 6 = 5x - 15$

$x = 7$

5min=1/60=1/12 hour
2min=2/60=1/30 hour

NOV 14

let $x = 56928$

$\sqrt{x^2 - (x+2)(x-2)}$

$= \sqrt{x^2 - x^2 + 4} = \sqrt{4} = 2$

NOV 13

$(ab)x = a(a+b)b$

$(10a+b)x = 100a + 10(a+b) + b$

$= 100a + 10a + 10b + b$

$x = \frac{110a + 11b}{10a+b} = 11$

NOV 15

$0.\overline{3}\left(\lim_{n\to\infty} \frac{3\left(1-\frac{6}{n^2}\right)}{1}\right) \rightarrow 0.\overline{3}\cdot(3) = 1$

NOV 16 $x = (1/2)54° = 27°$

NOV 17

$$\begin{pmatrix} 1 & 0 & 1 \\ -2 & 3 & -1 \\ 0 & 1 & -4 \end{pmatrix} = 1\begin{pmatrix} 3 & -1 \\ 1 & -4 \end{pmatrix} - 0\begin{pmatrix} -2 & -1 \\ 0 & -4 \end{pmatrix} + 1\begin{pmatrix} -2 & 3 \\ 0 & 1 \end{pmatrix} = 1(-12+1) + 0 + 1(-2) = -13$$

NOV 18

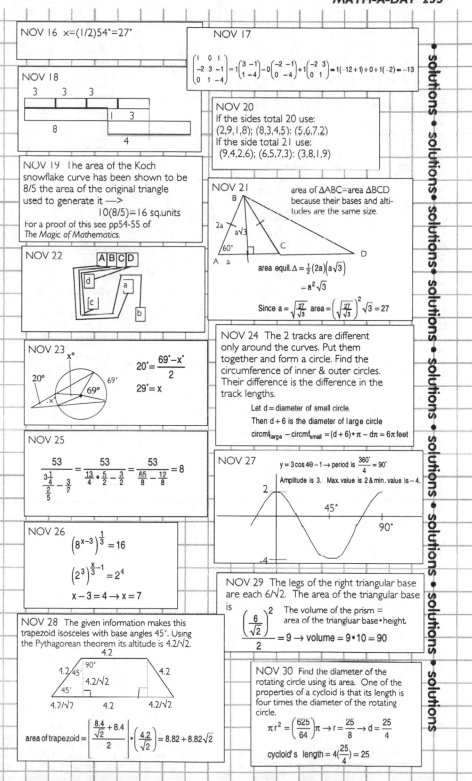

NOV 19 The area of the Koch snowflake curve has been shown to be 8/5 the area of the original triangle used to generate it —>

$10(8/5) = 16$ sq.units

For a proof of this see pp54-55 of *The Magic of Mathematics*.

NOV 20
If the sides total 20 use:
(2,9,1,8); (8,3,4,5); (5,6,7,2)
If the side total 21 use:
(9,4,2,6); (6,5,7,3); (3,8,1,9)

NOV 21

area of △ABC=area △BCD because their bases and altitudes are the same size.

area equil. $\triangle = \frac{1}{2}(2a)(a\sqrt{3})$

$= a^2\sqrt{3}$

Since $a = \sqrt{\frac{27}{\sqrt{3}}}$ area $= \left(\sqrt{\frac{27}{\sqrt{3}}}\right)^2 \sqrt{3} = 27$

NOV 22

NOV 23

$20° = \dfrac{69° - x°}{2}$

$29° = x$

NOV 24 The 2 tracks are different only around the curves. Put them together and form a circle. Find the circumference of inner & outer circles. Their difference is the difference in the track lengths.

Let d = diameter of small circle.
Then d + 6 is the diameter of large circle
circmf$_{large}$ − circmf$_{small}$ = (d+6)•π − dπ = 6π feet

NOV 25

$$\frac{53}{\frac{3\frac{1}{4}}{\frac{2}{5}} - \frac{3}{2}} = \frac{53}{\frac{13}{4} \cdot \frac{5}{2} - \frac{3}{2}} = \frac{53}{\frac{65}{8} - \frac{12}{8}} = 8$$

NOV 26

$$\left(8^{x-3}\right)^{\frac{1}{3}} = 16$$

$$\left(2^3\right)^{\frac{x}{3}-1} = 2^4$$

$$x - 3 = 4 \rightarrow x = 7$$

NOV 27

$y = 3\cos 4\theta - 1 \rightarrow$ period is $\dfrac{360°}{4} = 90°$

Amplitude is 3. Max. value is 2 & min. value is −4.

NOV 28 The given information makes this trapezoid isosceles with base angles 45°. Using the Pythagorean theorem its altitude is 4.2/√2.

area of trapezoid $= \left| \dfrac{\frac{8.4}{\sqrt{2}} + 8.4}{2} \right| \cdot \left(\dfrac{4.2}{\sqrt{2}}\right) = 8.82 + 8.82\sqrt{2}$

NOV 29 The legs of the right triangular base are each 6/√2. The area of the triangular base is

$$\dfrac{\left(\dfrac{6}{\sqrt{2}}\right)^2}{2} = 9 \rightarrow$$

The volume of the prism = area of the triangular base • height.

volume $= 9 • 10 = 90$

NOV 30 Find the diameter of the rotating circle using its area. One of the properties of a cycloid is that its length is four times the diameter of the rotating circle.

$$\pi r^2 = \left(\dfrac{625}{64}\right)\pi \rightarrow r = \dfrac{25}{8} \rightarrow d = \dfrac{25}{4}$$

cycloid's length $= 4\left(\dfrac{25}{4}\right) = 25$

DEC 1 There are 91.
1 square of size 6×6
4 squares of size 5×5
9 squares of size 4×4
16 squares of size 3×3
25 squares of size 2×2
36 squares of size 1×1

DEC 2

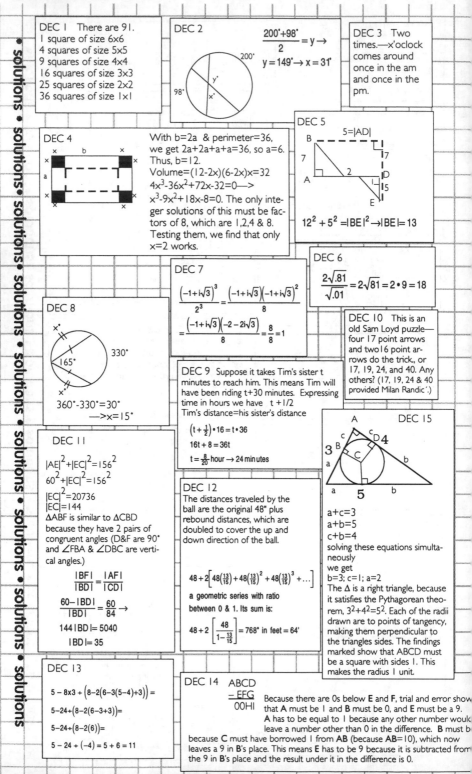

$$\frac{200°+98°}{2}=y \rightarrow$$
200°
$$y=149° \rightarrow x=31°$$
98°
y°
x°

DEC 3 Two times.—x'oclock comes around once in the am and once in the pm.

DEC 4

b
a
x x x x x x

With b=2a & perimeter=36, we get 2a+2a+a+a=36, so a=6. Thus, b=12.
Volume=(12-2x)(6-2x)x=32
$4x^3-36x^2+72x-32=0 \longrightarrow$
$x^3-9x^2+18x-8=0$. The only integer solutions of this must be factors of 8, which are 1,2,4 & 8. Testing them, we find that only x=2 works.

DEC 5
5=|AD|
B
7
7
A
2
D
15
E
$12^2 + 5^2 =|BE|^2 \rightarrow |BE|=13$

DEC 6
$$\frac{2\sqrt{.81}}{\sqrt{.01}}=2\sqrt{81}=2 \cdot 9=18$$

DEC 7
$$\frac{\left(-1+i\sqrt{3}\right)^3}{2^3}=\frac{\left(-1+i\sqrt{3}\right)\left(-1+i\sqrt{3}\right)^2}{8}$$
$$=\frac{\left(-1+i\sqrt{3}\right)\left(-2-2i\sqrt{3}\right)}{8}=\frac{8}{8}=1$$

DEC 8
x°
330°
165°
x°
$360°-330°=30°$
$\longrightarrow x=15°$

DEC 9 Suppose it takes Tim's sister t minutes to reach him. This means Tim will have been riding t+30 minutes. Expressing time in hours we have t +1/2
Tim's distance=his sister's distance
$$\left(t+\tfrac{1}{2}\right) \cdot 16 = t \cdot 36$$
$$16t + 8 = 36t$$
$$t=\tfrac{8}{20} \text{ hour} \rightarrow 24 \min \text{utes}$$

DEC 10 This is an old Sam Loyd puzzle—four 17 point arrows and two 16 point arrows do the trick., or 17, 19, 24, and 40. Any others? (17, 19, 24 & 40 provided Milan Randic´.)

DEC 11
$$|AE|^2+|EC|^2=156^2$$
$$60^2+|EC|^2=156^2$$
$$|EC|^2=20736$$
$$|EC|=144$$
ΔABF is similar to ΔCBD because they have 2 pairs of congruent angles (D&F are 90° and ∠FBA & ∠DBC are vertical angles.)
$$\frac{|BF|}{|BD|}=\frac{|AF|}{|CD|}$$
$$\frac{60-|BD|}{|BD|}=\frac{60}{84} \rightarrow$$
$$144|BD|=5040$$
$$|BD|=35$$

DEC 12
The distances traveled by the ball are the original 48" plus rebound distances, which are doubled to cover the up and down direction of the ball.
$$48+2\left[48\left(\tfrac{13}{15}\right)+48\left(\tfrac{13}{15}\right)^2+48\left(\tfrac{13}{15}\right)^3+...\right]$$
a geometric series with ratio between 0 & 1. Its sum is:
$$48+2\left[\frac{48}{1-\tfrac{13}{15}}\right]=768" \text{ in feet }=64'$$

DEC 15

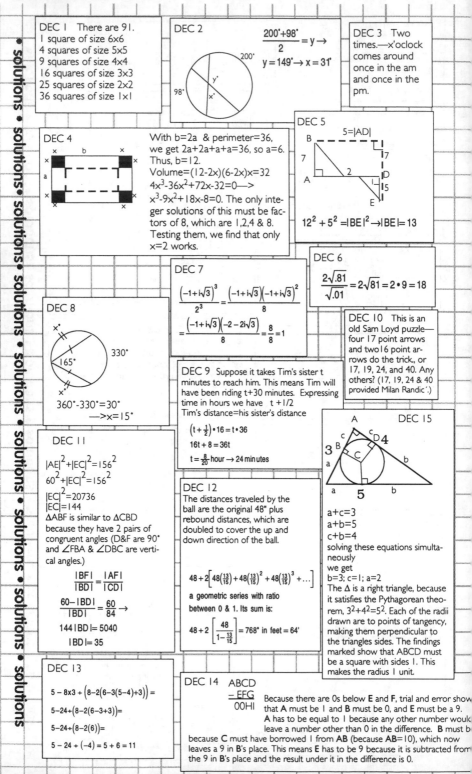

A
c
D 4
3 B
c
C
a
a
5
b
b
a+c=3
a+b=5
c+b=4
solving these equations simultaneously
we get
b=3; c=1; a=2
The Δ is a right triangle, because it satisfies the Pythagorean theorem, $3^2+4^2=5^2$. Each of the radii drawn are to points of tangency, making them perpendicular to the triangles sides. The findings marked show that ABCD must be a square with sides 1. This makes the radius 1 unit.

DEC 13
$$5-8\times3+\left(8-2\left(6-3(5-4)+3\right)\right)=$$
$$5-24+\left(8-2(6-3+3)\right)=$$
$$5-24+\left(8-2(6)\right)=$$
$$5-24+(-4)=5+6=11$$

DEC 14
  ABCD
− EFG
─────
  00HI

Because there are 0s below **E** and **F**, trial and error show that **A** must be 1 and **B** must be 0, and **E** must be a 9. **A** has to be equal to 1 because any other number would leave a number other than 0 in the difference. **B** must be because **C** must have borrowed 1 from **AB** (because **AB**=10), which now leaves a 9 in B's place. This means **E** has to be 9 because it is subtracted from the 9 in B's place and the result under it in the difference is 0.

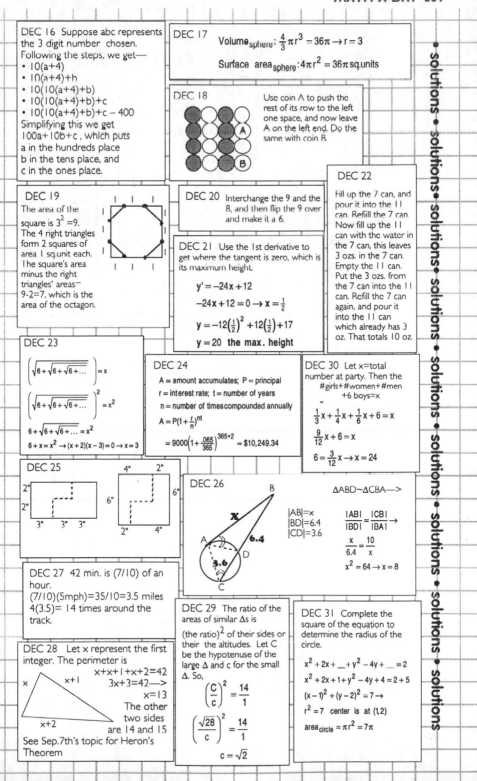

DEC 16 Suppose abc represents the 3 digit number chosen. Following the steps, we get—
- $10(a+4)$
- $10(a+4)+h$
- $10(10(a+4)+b)$
- $10(10(a+4)+b)+c$
- $10(10(a+4)+b)+c-400$

Simplifying this we get $100a+10b+c$, which puts a in the hundreds place b in the tens place, and c in the ones place.

DEC 17

$$\text{Volume}_{\text{sphere}}: \frac{4}{3}\pi r^3 = 36\pi \rightarrow r = 3$$

$$\text{Surface area}_{\text{sphere}}: 4\pi r^2 = 36\pi \text{ sq.units}$$

DEC 18 Use coin A to push the rest of its row to the left one space, and now leave A on the left end. Do the same with coin B.

DEC 19 The area of the square is $3^2 = 9$. The 4 right triangles form 2 squares of area 1 sq.unit each. The square's area minus the right triangles' areas— $9-2=7$, which is the area of the octagon.

DEC 20 Interchange the 9 and the 8, and then flip the 9 over and make it a 6.

DEC 21 Use the 1st derivative to get where the tangent is zero, which is its maximum height.

$$y' = -24x + 12$$
$$-24x + 12 = 0 \rightarrow x = \tfrac{1}{2}$$
$$y = -12\left(\tfrac{1}{2}\right)^2 + 12\left(\tfrac{1}{2}\right) + 17$$
$$y = 20 \text{ the max. height}$$

DEC 22 Fill up the 7 can, and pour it into the 11 can. Refiill the 7 can. Now fill up the 11 can with the water in the 7 can, this leaves 3 ozs. in the 7 can. Empty the 11 can. Put the 3 ozs. from the 7 can into the 11 can. Refill the 7 can again, and pour it into the 11 can which already has 3 oz. That totals 10 oz.

DEC 23

$$\sqrt{6+\sqrt{6+\sqrt{6+\ldots}}} = x$$

$$\left(\sqrt{6+\sqrt{6+\sqrt{6+\ldots}}}\right)^2 = x^2$$

$$6+\sqrt{6+\sqrt{6+\ldots}} = x^2$$

$$6+x = x^2 \rightarrow (x+2)(x-3) = 0 \rightarrow x = 3$$

DEC 24

A = amount accumulates; P = principal
r = interest rate; t = number of years
n = number of times compounded annually

$$A = P\left(1+\frac{r}{n}\right)^{nt}$$

$$= 9000\left(1+\frac{.065}{365}\right)^{365 \cdot 2} \approx \$10,249.34$$

DEC 30 Let x=total number at party. Then the #girls+#women+#men +6 boys=x

$$\tfrac{1}{3}x + \tfrac{1}{4}x + \tfrac{1}{6}x + 6 = x$$
$$\tfrac{9}{12}x + 6 = x$$
$$6 = \tfrac{3}{12}x \rightarrow x = 24$$

DEC 25

4"   2"
2"
2"
6"   6"
3" 3" 3"
2"   4"

DEC 26

$|AB| = x$
$|BD| = 6.4$
$|CD| = 3.6$

$\triangle ABD \sim \triangle CBA \longrightarrow$

$$\frac{|AB|}{|BD|} = \frac{|CB|}{|BA|} \rightarrow$$

$$\frac{x}{6.4} = \frac{10}{x}$$

$$x^2 = 64 \rightarrow x = 8$$

DEC 27 42 min. is (7/10) of an hour.
(7/10)(5mph)=35/10=3.5 miles
4(3.5)= 14 times around the track.

DEC 28 Let x represent the first integer. The perimeter is

$$x+x+1+x+2=42$$
$$3x+3=42 \longrightarrow$$
$$x=13$$

The other two sides are 14 and 15

See Sep.7th's topic for Heron's Theorem

DEC 29 The ratio of the areas of similar $\triangle$s is (the ratio)^2 of their sides or their altitudes. Let C be the hypotenuse of the large $\triangle$ and c for the small $\triangle$. So,

$$\left(\frac{C}{c}\right)^2 = \frac{14}{1}$$

$$\left(\frac{\sqrt{28}}{c}\right)^2 = \frac{14}{1}$$

$$c = \sqrt{2}$$

DEC 31 Complete the square of the equation to determine the radius of the circle.

$$x^2 + 2x + \_ + y^2 - 4y + \_ = 2$$
$$x^2 + 2x + 1 + y^2 - 4y + 4 = 2 + 5$$
$$(x-1)^2 + (y-2)^2 = 7 \rightarrow$$
$$r^2 = 7 \text{ center is at } (1,2)$$
$$\text{area}_{\text{circle}} = \pi r^2 = 7\pi$$

DATE:	TOPIC:
13, Aug	+ and − symbols
3, Sep	17-gon of Gauss
26, Feb	3-d representations & computer programming
30, Aug	99 mishap
27, May	abacus, Greek
26, May	abacus, Roman
9, Nov	Abel's manuscript
7, Nov	Abel, Niels & quintic proof
25, Mar	Agnesi & education
26, Mar	Agnesi & Joseph Lagrange
24, Mar	Agnesi & the French Academy
23, Mar	Agnesi & the witch of Agnesi
16, Jan	Agnesi's math book
18, Oct	algebra & Greek mathematics
13, Apr	amicable numbers
16, Apr	angle trisection problem & Wantzel
9, Jun	approximations of pi
19, Jul	Arabic East & West numerals
29, Oct	Arcadia
9, Feb	Archimedes & math works auctioned
16, Mar	Archimedes' area of a sphere
15, Jul	Archimedes' Sand Reckoner
11, Jan	Archimedes, the work of
13, Oct	Aryabhata & his book of verse called Aryabhatiya
19, Jun	Asian counterpart/destruction of the Library of Alexandria
6, May	atomic kilogram
5, Jan	Attic numerals
6, Jan	Babbage & his works
1, Jul	Babylonian records
20, Dec	Beal's conjecture & contest
9, Jan	Bernoulli family
11, Dec	Big Bang & singularities
18, May	bit and qubit
10, Aug	bits & bytes
12, Jan	BMI (body mass index)
7, Jan	book on astronomy & Bhramagupta
11, Mar	Boolean algebra
1, Sep	Brahmi numerals
18, Jun	Brain, first computer virus
7, Aug	c & the continuum hypothesis
14, Mar	CAEN
23, May	calculus, lineage of term
24, May	calculus, Yenri
5, Apr	Cantor's one-to-one correspondence
26, Jun	Cardano-Tartaglia cubic equation problem
6, Aug	cardinals vs ordinals
10, Nov	Cauchy's bad habit
18, Nov	Cauchy-Kovalevsky theorem

About the Author

Mathematics teacher and consultant Theoni Pappas received her B.A. from the University of California at Berkeley in 1966 and her M.A. from Stanford University in 1967. Pappas is committed to demystifying mathematics and to helping eliminate the elitism and fear often associated with it. In 2000 she received The Excellence in Achievement Award from the University of California Alumni Association.

Her books have been translated into Japanese, Finnish, Slovakian, Czech, Korean, Turkish, simplified and traditional Chinese, Portuguese, Italian, and Spanish.

Her innovative creations include   *The Mathematics Calendar, The Math-T-Shirt,   The Children's Mathematics Calendar, The Mathematics Engagement Calendar,* and *What Do You See?*—an optical illusion slide show with text.

In addition to *Math-A-Day,* Pappas is also the author of the following books: *Mathematics Appreciation, The Joy of Mathematics, Greek Cooking for Everyone, Math Talk, More Joy of Mathematics, Fractals, Googols and Other Mathematical Tales, The Magic of Mathematics, The Music of Reason, Mathematical Scandals, The Adventures of Penrose —The Mathematical Cat, Math for Kids & Other People Too!, Mathematical Footprints,* and *Math Stuff. The Further Adventures of Penrose—The Mathematical Cat* is her most recent book.